IK IT Job Interview Series

# Cracking the ECE Skills

IK IT Job Interview Series

# Cracking the ECE Skills

**Dr. Shriram K. Vasudevan**
*Assistant Professor (Selection Grade)*
Dept. of Computer Science and Engineering
Amrita School of Engineering, Amrita University
Coimbatore, Tamil Nadu, India

**Dr. C. Vivekanandan**
*Professor (EEE) and Vice Principal*
SNS College of Engineering
Coimbatore

**Priyanka A.**
Amrita School of Engineering
Kodigehalli Road
Bengaluru

**Dr. D. P. Kothari**
*Director- Research*
Gaikwad-Patil Group of Institutions
Wardha Road
Nagpur

**I.K. International Publishing House Pvt. Ltd.**

NEW DELHI

*Published by*

I.K. International Publishing House Pvt. Ltd.
4435-36/7, Ansari Road, Daryaganj
New Delhi–110 002 (India)
E-mail: info@ikinternational.com
Website: www.ikbooks.com

ISBN: 978-93-85909-53-5



Published by Krishan Makhijani for I.K. International Publishing House Pvt. Ltd., 4435-36/7, Ansari Road, Daryaganj, New Delhi–110 002 and Printed by Rekha Printers Pvt. Ltd., Okhla Industrial Area, Phase II, New Delhi–110 020.

# PREFACE

Objective type question entails a user to provide a response to a question whose answer is fixed. They have one potential answer and generally there is no room for opinions. This helps the users to recall the correct answer from memory thereby sharpening their skill set. Hence, objective type questions are preferred by most companies for selecting the right candidate. In this book, we have focused more on Electronics and Communication Engineering along with other closely related fields.

Electronics Engineering deals mostly with electrical circuit that involves both active and passive electrical components along with other interconnection technologies. The semi-conductor companies who are designing and manufacturing of semiconductor products would like to hire employees who are experts in Electronics Engineering and this book will be of highly useful to them.

Electrical Engineering on the other hand deals with the study and applications of electricity and electromagnetism. This includes multiple subfields like electronics, digital engineering, telecommunications, control systems, instrumentation and microelectronics. A wide range of industries are looking for engineers with electrical skill set for designing, developing, testing and supervising the deployment of electrical systems and electronic devices. This book will help them test their skill set in this particular field.

Electrical machines study electric motors and generators. They are called electromechanical energy converters which help in converting electricity into mechanical power or vice versa. These machines consume about 60% of the electricity produced. Successful persons in this field will be able to demonstrate a proven capability and experience in electrical machine design and development roles as well across different electrical industries. This book helps them in gaining knowledge and problem solving on interpreting the machine requirements when they work in collaboration with systems, hardware and mechanical engineers to generate and design the machine types.

Communication Engineering related questions discussed in this book will bring together all electrical engineering disciplines discussed above along with systems engineering knowledge to enhance telecommunication systems. Engineers who are strong in this area can apply for positions in the industries focusing on complex electronic switching systems, fiber optics cabling, copper wire telephone facilities, IP data systems and terrestrial radio link systems for traditional communications.

An antenna is an electrical device that helps in converting electric power into radio waves and vice versa. It has a radio transmitter or a receiver as part of its components. They are designed to transmit and receive radio waves in all directions. Antenna and RF design engineers are wanted in most countries and the expected skill set comprises creating and implementing antenna system, design base station antenna, etc. This book covers questions from this area as well.

Object detection systems like radar uses radio waves to find the range, angle and velocity of different objects that can be used to detect aircraft, ships, motor vehicles, weather formations and terrain. The modern uses of this concept are very much diversified including air traffic control, ocean surveillance systems, flight control systems, missile target locating systems, etc., to name a few. The job opportunities are also hence very much diversified and this book would help the readers to gain sufficient knowledge on this area and crack the interviews.

Computer networks allow computers to exchange data between systems. A data link is used for this purpose. They differ in terms of medium through which the data travel, the protocols used in data exchange, the size of the network and organizational intent. We touch upon this area as well as it supports enormous number of applications such as access to the World Wide Web, digital audio-video broadcasting, printers and fax machines, etc., to name a few.

Embedded systems is the next major topic discussed in this book with ample amount of objective type questions. It is a computer system with a dedicated functionality defined to achieve within a larger electrical system often with real time constraints to be achieved. They are often based on microcontrollers and a special focus is given to this area as well. They range from portable devices including MP3 players, smart watches to larger installations like traffic lights and complex systems like avionics. They are also found in a wide variety of domains like consumer, cooking, automotive, industrial, military and medical applications. Every industry in the aforementioned areas will look to hire engineers who are strong in embedded systems concepts and debugging skills and this book will help them a lot to fine-tune their skill set in this area.

Basic understanding of electrical, electronics, computer networking, communications, embedded systems, microprocessors, microcontrollers, antennas and radars is a prerequisite to this book. This book will help them to further analyze and understand the concepts of these areas.

In order to help the end user of this book acquire knowledge on different areas surrounding the Electronics Engineering, this book is organized systematically. For each chapter the questions along with multiple choice answers along with the right answer and the explanation are detailed which makes it stand out from rest of the publications in similar areas. Each question has been marked with Level of difficulty identifier. L1 to L4 is the marking and L1 is basic whereas L4 would really test the reader's skill in the subject area.

The authors are from 4 different domains. One is Computer Science and Embedded System expert, two are electrical and electronics experts and one more from the Communication Engineering.

The break-up details of questions chapter-wise is detailed below:

| S. No | Title | Number of questions |
|---|---|---|
| 1. | Microprocessors, Microcontrollers and Embedded Systems | 183 |
| 2. | Digital Electronics | 149 |
| 3. | Communication Engineering | 106 |
| 4. | Operating Systems | 115 |
| 5. | C Programming | 150 |
| 6. | C++ | 145 |
| 7. | Electrical Machines | 105 |
| 8. | Electricals and Electronics | 55 |
| 9. | Antennas and Radars | 32 |
| 10. | Mechanical and Electronics | 9 |
| 11 | Computer Networks | 114 |

Total Questions = 1163

**Dr. Shriram K. Vasudevan**
**Dr. C. Vivekanandan**
**Priyanka A.**
**Dr. D. P. Kothari**

# CONTENTS

# 1 MICROPROCESSORS, MICROCONTROLLERS AND EMBEDDED SYSTEMS

This section covers questions from the areas of microprocessor, microcontroller and embedded systems. Most of the questions are theoretical and would not require the candidates to answer more than 1 minute. Students are requested not to memorize anything. Instead, if they understand, it would be easier.

1. **What is the difference between a microprocessor and a microcontroller? (L2)**

   Everything on board is called a microcontroller and everything off the board is the microprocessor. In short, a microcontroller by itself is a microcomputer, whereas a microprocessor is not. It would need peripherals to get qualified as a microcomputer. Also, a microcontroller is meant to perform single function. But, a microprocessor is not. One can understand the difference from construction point of view from the diagrammatic representation given below.

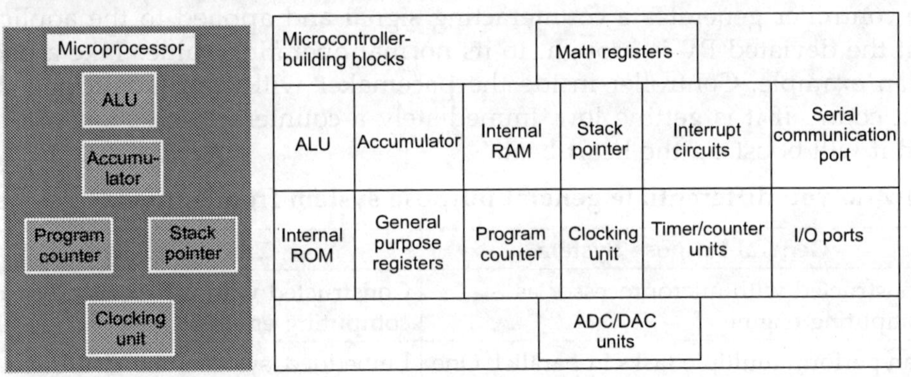


**Fig. 1.1** Microprocessor and microcontroller (LHS and RHS respectively)

2. **When would someone opt for a microprocessor instead of a microcontroller for building a system? (L2)**

   When the system to be built is single purpose system, i.e., like washing machine or air conditioner, which would not do anything other than washing and cooling, microcontroller is preferred. If an equipment is designed to handle multiple tasks, then microprocessor is preferred.

3. **What are the typical characteristics of embedded systems? (L2)**

   Any embedded system would have the following characteristics:

- ➢ Single functioned
- ➢ Tightly constrained (power, size and cost)
- ➢ Real time and reactive
- ➢ Easier to use and got interactive user interface

4. **What are the major challenges that one would face when building an embedded system? (L2)**

   While building an embedded system, the developer tends to face a lot of challenges. Few of them are listed below:

   - ➢ How much emphasise should be given to software/hardware?
   - ➢ How to control and reduce power consumption?
   - ➢ How to make it easier for future upgrade and repairs?
   - ➢ How to test it?
   - ➢ How to meet the deadlines both in terms of the product release and the product output.

5. **What is an embedded system? (L2)**

   Many say, anything which has software and hardware together is called an embedded system. Few say, anything other than desktop system is an embedded system. One can define an embedded system in a better way as "an electronic controller built into the application, continuously monitoring the process variables and ensures that the process variable (PV) does not change; in the event of a change the controller generates a counteracting signal and applied to the application so that the deviated PV is brought to its normal operating value. Take a pacemaker as an example. Controller inside the pacemaker will keep monitoring the heart beat count. If it is getting low, immediately a counteracting action will be taken and it will boost up the heart beat."

6. **How do you differentiate general purpose system from embedded system? (L3)**

| General-Purpose System | Embedded System |
|---|---|
| Constructed with microprocessor as computing engine. | Constructed with microcontroller as computing engine. |
| Can perform multiple tasks in parallel. One example would be Laptop or a desktop computer. One can play music while using a compiler to program a logic with internet being accessed for reference. | Embedded systems are dedicated for one specific task. For example, the water heater which is built with a microcontroller cannot be used to play music. It may be impossible for embedded systems to be designed to do multiple tasks. |
| Deadlines are not so crucial, i.e., one may not be hurt if the output of the system is delayed, slightly. | Mostly deadlines are crucial. Hence, they are referred as real time systems. If deadlines are crossed, it could prove fatal. (Example: car braking system). |
| Example operating systems: Windows XP, Ubuntu, etc. | Example operating systems: VxWorks, WinCE, RT Linux, etc. |

**7. Can someone say, a microcontroller has a microprocessor inside and vice versa is not true? (L1)**

Yes. Very much.

**8. What are the major components of a microprocessor/microcontroller? (L2)**

Following are the important components:

(a) ALU (Arithmetic and Logical Unit)
(b) Control Unit.
(c) Timer Unit.
(d) Memory (RAM and ROM).
(e) Stack pointer register.
(f) Program counter register.
(g) Flag registers.
(h) Bus.

**9. What is a register made of? (L2)**

All the registers are constructed with flip-flops. Mostly D-Flip-Flop is the one used for constructing the register. A simple 8-bit register looks something like shown below where 8 flip-flops are connected to get 8-bit register constructed.

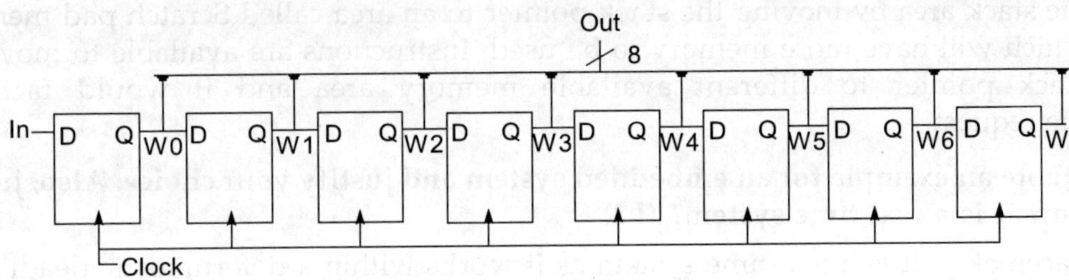


**Fig. 1.2** An eight-bit register.

**10. What is a program counter made of? How is it useful? Is it mandatory for any microprocessor or microcontroller? (L2)**

A program counter is also a register and hence it is made of flip-flops. The width of program counter is different for different processors. It could be an 8-bit program counter or a 16-bit program counter based on the architecture of the processor. A program counter register is used to hold/store the address of the next instruction to be executed. So it saves time and execution can be made faster, i.e., pipelining. It is mandatory for any processor to have a program counter.

**11. Can someone access the program counter? If yes, what is the address in which program counter will reside? (L3)**

Program counter cannot be accessed by the programmer as it would lead to distress in the code execution if the flow is altered. Hence, the manufacturers do not provide the address of program counter to the end users thereby preventing them from accessing it.

**12. As program counter, there is another register called stack pointer register, which is equally respected. Why is it so? (L3)**

Stack is a temporary storage area where the values can be stored. For an instance, when a function is called like add (3,4), the arguments 3, 4 would be kept in the stack (the operation is called pushing) and it is a temporary storage area. Once the function call is done and result is ready, the result will be again kept on the top of the stack. Once the execution is complete, the stack can be cleaned (the operation is referred to as popping). Stack pointer is a register used to point the stack.

**13. How big or small would be the stack? Can the programmer increase the size of the stack if need be? (L3)**

While the processor is sold, the size of the stack would be fixed. Say, something like 128 MB. But, the programmer would always be given the comfort of expanding the stack area by moving the stack pointer to an area called Scratch pad memory, which will have more memory to be used. Instructions are available to move the stack pointer to different available memory area and it would facilitate the request.

**14. Quote an example for an embedded system and justify your choice. Also, justify how it is a real time system? (L3)**

Pacemaker. It is a real-time system as it works within a deterministic deadline to protect the patient if the heartbeat goes low. If the pacemaker does not work on time, then the patient may die.

**15. Can someone call laptop or a desktop system an embedded system? Justify. (L2)**

No. They are general purpose systems and can do multiple tasks. Also, a laptop or desktop computer has a microprocessor inside and not a microcontroller.

**16. While building a system with a microprocessor, one cannot blindly select the microprocessor. What are the considerations to be made for selecting a microprocessor? (L3)**

The following are the most important parameters to be looked into for the selection of microprocessor or microcontroller:

- ➤ The most important factor is "number of transistors on the chip". One can easily see that the number of transistors are increasing drastically over years based on Moore's law. More the number of transistors, better the prospects of the processor.

- Micron is a factor. A micron is the width, in microns, of the smallest wire on the chip. To understand in an easier way, a human hair is 100 microns thick.
- Everyone wants the microprocessor to process the input faster and give results faster. Clock speed decides the speed of the chip. So, while selecting a processor, one should also see the operating speed of the processor. 8085 has 3 MHz while modern-day processors are in terms of GHz.
- Data width is the width of ALU, in the sense width of the data that ALU can process. It can be 8-bit, 16-bit or 32-bit. Based on the requirement, one would select 8-bit or 16-bit or 32-bit processor.
- MIPS is million instructions per second. It is the number of instructions that the processor can execute in a second. One can see that MIPS count has consistently increased and more the MIPS, better the processing speed.
- Cost is also an important factor to be considered which would in fact increase the overall cost of the system.

**17. What exactly is the role of the bus in the microprocessor or microcontroller? (L3)**

As the term says, bus is the one which can carry information from one point to another. It is like a typical bus that carries people from source to destination. Instead of people, here data is taken from source to destination. Without bus, no data can go in or go out of/from microprocessor or microcontroller.

**18. What is ALE and how is it useful in 8085 microprocessor or any other processor which has it? (L4)**

ALE is expanded as Address Latch Enable. It is useful when the address is placed on the address lines AD0–AD7. It is a simple pulse. This is deployed for the purpose of enabling the latch to save the address bits from the address lines and through that, address bus and data buses are de-multiplexed clearly. Means, the AD0–AD7 are capable of holding address or data. The manufacturers have reduced the number of pins in the processor by making AD0–AD7 flexible and versatile. To tell if the pin gets address or data, ALE is used. Not only for 8085, but also for many other processors ALE is used.

**19. Is there any register that programmers cannot use for storing data? For instance, like 10 H cannot be stored in this particular register. (L3)**

There are registers called Special Function Registers (SFRs) which would not enable the user to use it for general purpose, i.e., for storing the data of user's wish. Program counter can be cited as an example where user cannot store data 10 H. It can store only the address, that too automatically and no user intervention is required.

**20. How is an accumulator useful in the microprocessor? (L3)**

Every microprocessor or microcontroller would have something called Math Register. They are inevitable to be used to perform any math operation. For instance to add two numbers, Accumulator register must have one of the operands while the other operand can be stored in other general purpose register. Coming to 8085 or 8051 Accumulator is referred by letter "A" and is a math register. When there is a new 8-bit data entering, the previously stored data will get automatically over-written. Any operation that is happening will happen through Accumulator register only. Figure shown below shows how a register would be connecting to ALU. The status of the operations is updated in flag register.

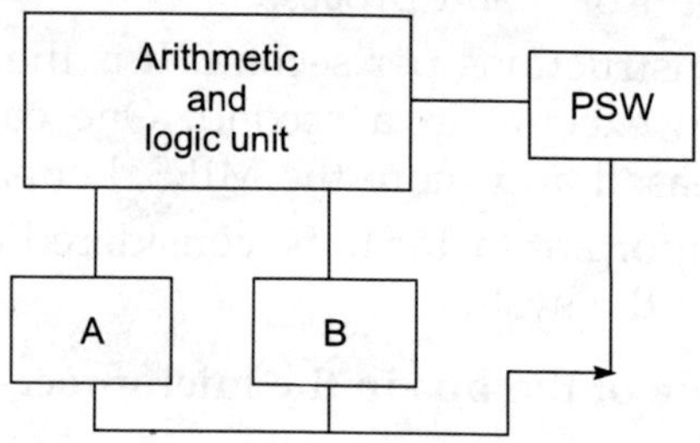


A = Accumulator.
B = Math register.
PSW = Flag register.

**Fig. 1.3** Accumulator with ALU

**21. How important is general purpose register (GPR) in a microprocessor or a controller? What general purpose registers are available in the 8085? (L2)**

As the name specifies, it can be used by programmers as they wish. GPRs are not like Special Function Registers (SFRs). GPRs can be used for all arithmetic and logical operations as per programmers' wish. But certain special function registers cannot be used by programmers as per the way they want. Here, in 8085 there are 6 general purpose registers. They are:

- B
- C
- D
- E
- H and
- L

Similarly, in 8051 there are 8 registers each in 4 banks counting to 32 registers totally as general purpose registers.

**22. How 8-bit general purpose registers can be used for handling 16-bit data? (L2)**

Most of the 8-bit processors and microcontrollers provide the programmer with an option of combining two 8-bit registers to handle the 16-bit data. In 8085 one can do this using the right pair as BC, DE and HL. Similar options are available in other microcontrollers too.

**23. What is the difference between Von-Neumann and Harvard architectures? (L3)**

Harvard architecture has separate memories for program and data while same memory is used for data and program in Von-Neumann architecture. The former is preferred over the latter because of bandwidth related advantages.

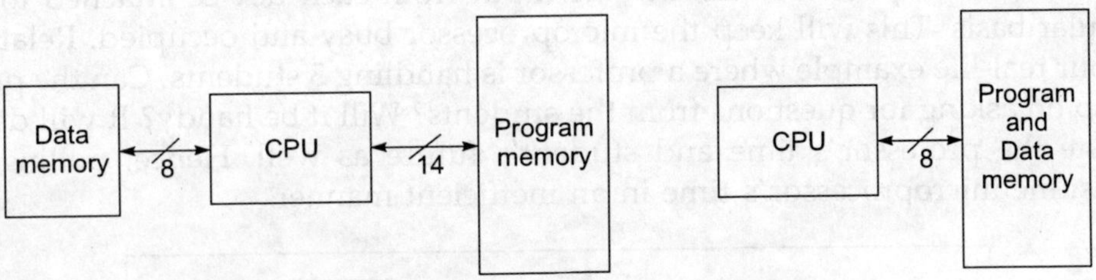


**Fig. 1.4** Harvard and Von-Neumann architecture.

**24. Any microprocessor would have buses inside for the data and instruction to be handled. Elaborate the types of buses with reference to 8085. (L3)**

There are three bus categories available in 8085 and they are:

- Address bus
- Data bus
- Control bus.

**1. Address bus:** As the name indicates it is used to carry the address. 8085 has 16 address lines which means it can have 2^16 = 65536 byte memory locations. The address bus is mainly used to recognize a memory location or a connected peripheral. Postman basically delivers the letters using address. Likewise it is mandatory to have the identity for the memory locations and it is referred to be as an address. Address bus is always unidirectional. The communication happens from the microprocessor to the peripherals.

**2. Data bus:** Data buses are bidirectional wires which carry data from or to microprocessor. Any information that gets in or goes out of microprocessor is through the data bus only. In 8085, there are 8 data lines from D0 to D7, and this is the reason why we call it an 8-bit processor. Data bus is used to carry the instructions, results of the operations, etc., to the peripherals and memory unit. It is bidirectional. When we quote that it is an 8-bit processor, it implies that the large data chunk has to be broken into smaller ones of up to 255 (0 to 255, 28).

**3. Control bus:** As the name suggests, it is very obvious. All the controlling actions are carried out through these lines. It can be unidirectional or bidirectional. The control signals will intimate the microprocessor on where to read or where to write. Many control signals are available with 8085.

**25. What is polling? Explain with an example. (L4)**

Assuming a professor is teaching in a class. He has to deal with 3 students in that class. How can it be done? Relate this to a situation where a microprocessor or a microcontroller has to handle 3 devices as shown in figure below. How can this be done? Two ways are there. First method is polling and the same is explained here. The microprocessor can ask for input from each device attached to it on a regular basis. This will keep the microprocessor busy and occupied. Relating this to our real-life example where a professor is handling 3 students. Can the professor keep on asking for questions from the students? Will it be handy? It will definitely waste the professor's time and student's future as well. Hence, polling would consume microprocessor's time in an inefficient manner.

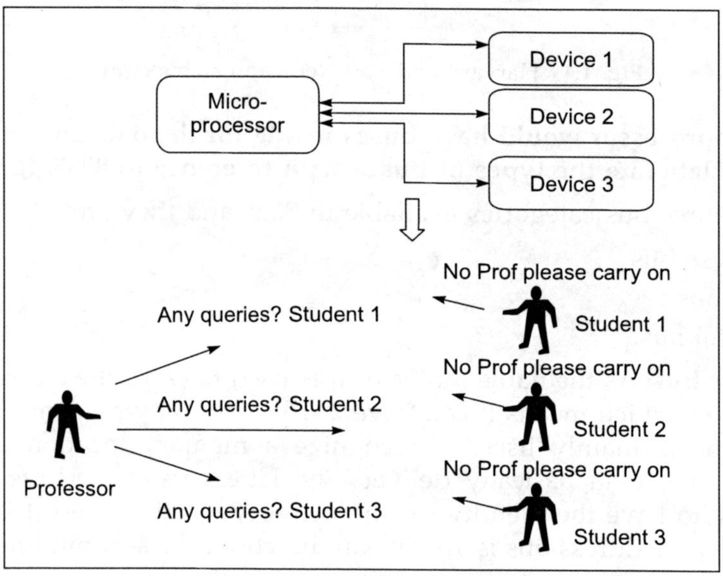


**Fig. 1.5** Polling. A microprocessor talking to peripherals.

**26. What is an interrupt? Explain with an example. (L4)**

Keeping the example shown in Fig. 1.5, one can answer this question too. Whenever a device needs the service of a microprocessor it will ask, i.e., peripherals will knock the door of the processor, when needed. Similarly, when any of the students gets a query, it will be raised then and there. Otherwise the professor will continue teaching, i.e., microprocessor can do something else other than waiting for the devices to answer. Figure in the previous answer is redrawn as the one shown below.

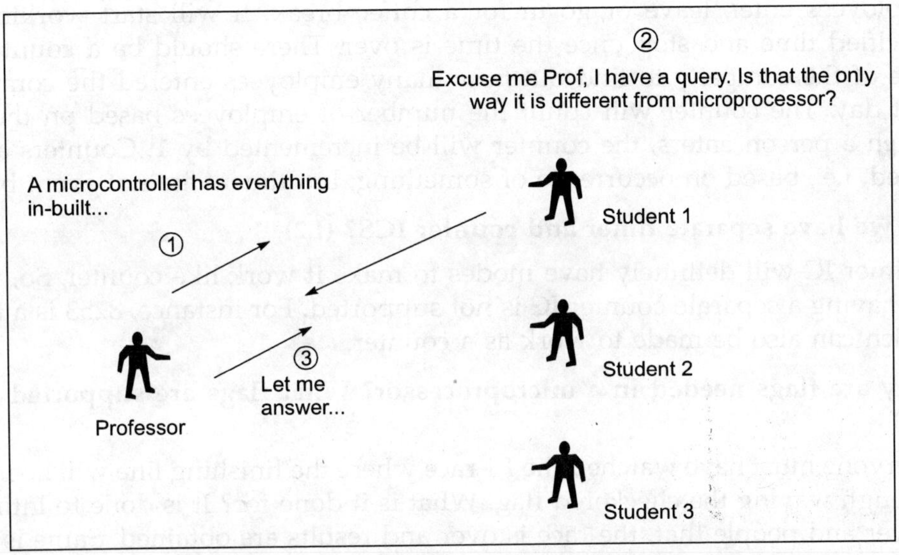


**Fig. 1.6** Interrupts. A microprocessor talking to peripherals.

**27. Which one would you prefer among Interrupt and Polling? (L3)**

The preference is definitely interrupts as the microprocessor can do other high priority tasks instead of wasting time polling for service requests.

**28. When interrupts are said to be efficient, why polling still persists? (L3)**

This is a valid question. There are scenarios where interrupts cannot be used, for example, if there is an accident, an interrupt could be raised and air bag could be released. There are monitoring systems in the modern cars which tests the air pressure of the tyres, cooling levels and other important parameters periodically. This is done on the regular basis and no interrupt is required here. Polling will be useful here. Microprocessor will poll for inputs from all the peripherals attached for analysis, periodically.

**29. Why do designers add additional timers in the motherboard when there is a timer available in the processor itself? (L4)**

Well, this is a tricky technical question. Though most of the microprocessors or microcontrollers have timers inside, designers keep additional timer ICs in the board. This is to prevent microprocessor from doing simple timer job and instead it can focus on something of higher priority and importance. Just to utilize the microprocessor or microcontroller more effectively, one would go in for additional chip, though it increases the cost.

**30. How is a timer different from a counter? (L4)**

This is a frequently asked question in most of the interviews. Let's take an example. Employees in a production unit have to work for 9 hours and the timer would be started by 9.00 AM and would be stopped after 9 hours. It would not mind if any employees enter, leave or go in for a coffee break. It will start working at the specified time and stop once the time is over. There should be a counter in the gate of the company to find out how many employees entered the company for that day. The counter will count the number of employees based on their entry. When a person enters, the counter will be incremented by 1. Counters are event based, i.e., based on occurrence of something, but timers are not event based.

**31. Do we have separate timer and counter ICS? (L2)**

A timer IC will definitely have modes to make it work like counter. So, the need for having a separate counter IC is not supported. For instance, 8253 is a timer IC, which can also be made to work as a counter.

**32. Why are flags needed in a microprocessor? What flags are supported in 8085? (L4)**

Everyone must have watched the F1 race where the finishing line will be intimated through waving the checkered flag. What is it done for? It is done to intimate the driver and people that, the race is over and results are obtained. Same is the case with any processor.

**Fig. 1.7** Flag used to indicate the race results.

When an operation is carried out by a microprocessor or a microcontroller, the results are not only stored somewhere, but also the status of the operations can be known through the flags. Every microprocessor has flags to indicate if the result is negative, zero or there was a carry, etc. the 8085 has five flip-flops acting as the flag registers. These can be set or reset based on the result of the operation being performed. The flags are Zero, Carry. Sign, Parity and Auxiliary carry flags. Based on the result obtained after the operations, the flags will be set accordingly. For example, if an addition operation results in a carry, CY flag will be set. Similarly, when an arithmetic operation results in a zero, it will be indicated by setting the ZERO flag. Similarly, when there is even number of 1s in the result, even parity will be set and it can be indicated by parity flag.

**33. What are the two types of interrupts supported in 8085? How are they different from each other? (L3)**

There are two types of interrupts supported in 8085. They are maskable and non-maskable interrupts. Through instructions if someone can disable the interrupts, then it is referred to as maskable interrupts. If an interrupt cannot be disabled through instructions, then it is said to be non-maskable.

**34. Which is the high priority non-maskable interrupt in 8085? (L2)**

TRAP carries a very high priority and it cannot be stopped by any command or any other mask. It is of the highest priority among all the other interrupts.

**35. What is an ISR (Interrupt Service Routine)? (L2)**

When an interrupt is raised or received, the corresponding piece of code to perform the desired action for the interrupt will be called and executed. They are not similar to function calls, as ISRs do much more than simple function call. While executing code for the interrupt which was raised, the processor should make sure that the current program counter is stored, current context and results are stored and then the ISR execution starts.

**36. What is context switching? Give an example. (L3)**

When the processor is executing a process, the process control block along with program counter, CPU registers, stack pointer, processor status register will be available with the processor. This is called the context of the program. When the processor wants to leave this process and execute another process, the corresponding PCB (process control block) and other information must be loaded. So the context of the current process will be stored in a good condition, only then the new context will be loaded in place. Changing contexts from one process to another is called context-switching.

**37. We normally use a term main memory frequently in OS/embedded systems. What is main memory? Is there any other memory available in the system? (L2)**

Main memory is the primary memory of the system. It is the Random Access Memory (RAM). It consists of many fast registers, directly connected to the processing unit. For a program to execute, it must be present in this main memory. So it will be loaded on to the RAM when launched. Also, this main memory is a temporary memory, i.e., its contents will be deleted when the power supply is stopped. Apart from main memory, there are secondary memory and removable/ extendable memory. Secondary memory is the hard disk and the permanent memory. Extendable memories are like pen drives, CDs, hard disk, etc. They can be used by the system and can be removed from it.

**38. Mention the types of interrupts that you are aware of. (L3)**

Based on the mode of interrupt, it is classified as hardware and software interrupts. Hardware interrupts are generated physically like, buttons, signals, IR, sensor outputs, etc. Software interrupts are generated using some simple codes. Also there is a class for raising software interrupts which is "Exception". They are unplanned interrupts. Exceptions are not manually generated like the others, it is generated automatically due to some flaw in the program/process execution. Also based on the frequency of occurrence interrupts are classified as periodic and aperiodic interrupts. If an interrupt occurs in a fixed interval of time, it is periodic. Else it occurs in any random point of time. This is called aperiodic interrupt.

**39. Define interrupt latency? (L3)**

When an interrupt occurs, the processor has to switch from the current process context to the ISR to respond to the interrupt. The total time taken to respond to the interrupt is called interrupt latency. Generally, it is the sum of three components. Switching time which is the time taken for context switching from process to interrupt, disable time which is the time taken to store and disable the currently executing file, and finally, execution time, the time taken to execute some critical section and higher priority interrupts. Lesser the latency, better the performance would be.

Latency = Switching time + Disable time + Execution time.

**40. What is an instruction? How is opcode and operand related to an instruction? (L2)**

Instruction is nothing but a command which makes the microprocessor perform a specific function. All the instructions are collectively grouped as instruction set. An instruction is basically a combination of two parts; one is operation code and the other is operands. Operation code is also called opcode. Operands can be one or two based on the instruction. The opcode will be used to identify on what kind of operation needs to be performed like, addition or subtraction or data movement and so on. And operands will be used to spot the source and destination. An operand can be any of the following:

- Registers (like accumulator, B register and so on)
- Immediate value (say 10H, 010101B)
- An address, i.e., memory location (say 30H)
- Any of the supported input or output ports.

**41. What is an addressing mode? Why is it important? (L2)**

The CPU can fetch information and access data in multiple ways. Generally, the operands are vital and can be used to fetch information and process it. The ways followed to access the data are called addressing modes.

**42. What are the addressing modes supported in 8085? (L2)**

1. Immediate Addressing
2. Direct Addressing
3. Register Addressing
4. Register Indirect Addressing
5. Implicit Addressing

Not only 8085, but all other processors have similar addressing modes.

**43. State an example for immediate addressing mode in 8085? How is immediate addressing done in 8051? (L2)**

This is the simplest mode of all modes, where the instruction itself will have the information to be moved or to be processed. An example is presented below. MVI A, 05H – 05H is the data to be moved to accumulator with MVI instruction which is abbreviation of Move Immediate. 8051 identifies immediate data through #. MOV A, #55H is an instance. There is no instruction called MVI in 8051 and hence is the arrangement.

**44. What is indirect addressing mode? State an example. (L2)**

In this mode, the place where the address is available will be specified. An example will clarify this better. LDAX D – Load the accumulator with the contents of the memory location whose address is stored in the register pair DE. The content will be available in the address pointed by DE and this is what is referred to be as indirect.

**45. State the purpose of instruction decoder in 8085? (L3)**

Instruction register stores the current instruction being executed while program counter stores the address of the subsequent instruction. Instruction decoder takes the instruction from the instruction register and then decodes the same. Decoded instruction is then moved to the next stage.

**46. Spot any 4 differences between RISC and CISC. (L3)**

| CISC | RISC |
|---|---|
| 1. Emphasis on hardware | Emphasis on software |
| 2. Multiple instruction sizes and formals | Instructions of same set with few formats |
| 3. Less registers | Uses more registers |
| 4. More addressing modes | Fewer addressing modes |

**47. State any one application where microprocessors are deployed. (L1)**

Traffic signals – Simple and effective example. The same can as well be done with a microcontroller.

**48. Mention whether the restart interrupts are supported by 8085? (L3)**

There are three signals present in this category. They are

- RST 7.5 (high priority)
- RST 6.5
- RST 5.5 (low priority)

Out of which RST 7.5 has the highest priority and RST 5.5 has the lowest priority.

**49. What is need for using HOLD signal in 8085? (L4)**

This is a very important and highly appreciable signal in 8085. If a process is using the data and address lines, and in case there is another process which is of much higher priority needing the address lines, HOLD signal can be used. When HOLD request is raised, the current process using the buses will release it as early as possible. Once the processing is complete, the process which was using it earlier can regain the buses.

**50. What is the need of HLDA signal in 8085? (L3)**

When a HOLD request is raised, there has to be an acknowledgement. HLDA is the acknowledgement signal raised in response to that. When the HOLD request is removed, HLDA will go low. CPU can then have control over the buses and proceed with normal flow.

**51. What is 8255 and how is it useful? (L3)**

8255 is a PPI and is expanded as Programmable Peripheral Interface. It is the most commonly used I/O chip which can be used with any microprocessor. It is useful in data transfer across the peripherals. 8085 gets connected to any peripheral through 8255. It interfaces 8085 to the outside world.

**52. What are the modes of operation supported in 8255? (L2)**

It supports 2 modes. I/O mode and BSR (Bit Set Reset) mode.

**53. What is 8259 and what for it used? (L2)**

8259 is referred to be as programmable interrupt controller and it is used to handle interrupts when the in-built support interrupts are not sufficient in 8085. Modern-day processors may not require these chips as all the capacity is built-in.

**54. Where is 8253/8254 deployed and what is it? (L2)**

It is referred to be as timer and it can be used for generation of time delay as a timer or counter operation. Many modes are available to generate desired output and 8254 is an advanced version of 8253. In the olden days, all the timer related operations of 8085 were done by having 8253 in place.

**55. What is the use of 8279 IC? (L2)**

It is used for interfacing the microprocessor 8085 with keyboard. It does most of the processing and relieves the processor from the load.

**56. How many register banks are available in 8051? (L2)**

4 register banks with each bank having eight registers each. Registers are named as R0–R7.

**57. What is the instruction used to select register bank 2 in 8051? (L2)**

SETB PSW.4, where PSW is the flag register for 8051 microcontroller. It is expanded as programmable status word.

**58. What will be the default value set in PSW.3 and PSW.4 upon reset in 8051? (L2)**

Bank 3.

**59. What is PSW expanded as in 8051? (L2)**

Program Status Word.

**60. What is the use of PSW in 8051? How does it look like? (L4)**

Microcontrollers have a provision for this and they are called FLAGS. There are few flags which help the user to know if the operations are happening as expected. And all these flags are accommodated in a special function register called program status word. And it is bit accessible. Figure 1.8 shows the PSW format

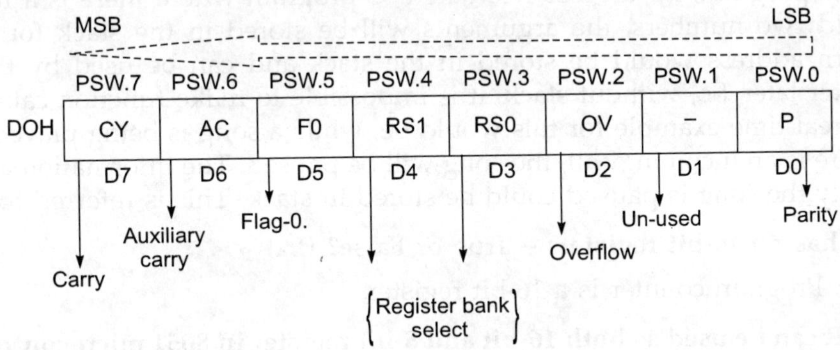


**Fig. 1.8** PSW register.

where,

- P – Parity flag – PSW.0 – it will be set or reset based on the value (numeric) present in the accumulator. If it has odd number of 1s then it is set and otherwise it is set to 0. So it serves just as an indicator.
- PSW.1 – This field is not used and reserved for future usage.
- OV – Overflow flag – PSW.2 – When result of a signed operation is very large, OV flag is used. It will be set in that case. Else will remain unset.

- RS0 and RS1 – PSW.3 and PSW.4 – Register Bank Select. There are four register banks supported in 8051 and each of them has 8 registers from R0 to R7. The user has a flexibility of selecting one of those banks using the register bank select options.

**61. How can one access accumulator without using name A, i.e., through address in the microcontroller 8051? (L2)**

0E0h is the address of Accumulator A with which one can access.

**62. What is the major drawback with respect to multiplication and division operation in 8085? (L2)**

Simple, there is no explicit instruction available to perform multiplication or division. Hence, multiplication is performed through repeated addition and division is performed through repeated subtraction. This is the major setback.

**63. Can division be done without B register in 8051? (L2)**

No. Math registers A and B are available and they are to be used for this task.

**64. When does auxiliary carry bit help in any microprocessor or a microcontroller? (L2)**

In BCD operations.

**65. State one application where stack is used frequently. (L2)**

In any program, for an instance take a C program where there is a function call to add two numbers, the arguments will be stored in the stack for processing. Return address would be stored in the stack and can be used by the program counter later. So, without stack, it is impossible to make function calls efficiently. One real-time example for this would be, while a song is being played in mobile, if there is an incoming call, the song will be paused. The information about where exactly the song is paused could be stored in stack. This is referred to as context.

**66. 8051 has no 16-bit registers – True or False? (L2)**

False. Program counter is a 16-bit register.

**67. DPTR can be used as both 16-bit and 8-bit register in 8051 microcontroller – True or False? (L2)**

True

**68. Upon reset, what is the value in the SP (stack pointer) register in 8051? (L2)**

07H, default register bank structure enables this way.

**69. Upon PUSHing data onto the stack, the SP register is ____________ (L2)**

Incremented.

**70. Upon POPing data from stack, the SP register is ___________ (L2)**

Decremented.

**71. The stack uses the same area of RAM as ___________ in 8051 microcontroller. (L2)**

Bank 1.

**72. How will someone be able to use bank 1 registers for general purpose storage in 8051? (L2)**

By moving the SP to point to different area, preferably to the area of scratch pad memory.

**73. Can someone make the SP pointer to point different memory area in 8051? If yes, how? (L2)**

Yes with MOV SP, # ADDRESS

**74. What is the RAM address of ACC and B in 8051? (L2)**

E0H and F0H. E0H is the address of the accumulator and F0H is the address of the math register B.

**75. Can programmer use program counter for general purpose storage? If yes, how? [L3]**

PC will hold the address of next instruction to be executed. No, it can never be used by a programmer for general purpose storage.

**76. Is "A" register bit addressable in 8051, which happens to be the accumulator? (L2)**

Yes. It is bit addressable.

**77. MOV DPTR, #676556; is this correct with respect to 8051? (L2)**

No. This is not the right way to use this instruction.

**78. How many pins are there in 8051 chip? (L2)**

40.

**79. What will happen when 9th pin in 8051 is set to 1? (L2)**

It is a reset pin and will reset all the registers.

**80. After reset where will the PC point to in 8051? (L2)**

It will be pointing to 0000 H. This is true with most of the microprocessors and microcontrollers too.

**81. For what 18th and 19th pins are meant in 8051? (L2)**

XTAL 1 and XTAL 2.

**82. INTEL is the only manufacturer of 8051. True or False (L2)**

False.

**83. Can instruction set prescribed for 8051 be used with 8052? (L2)**

Yes. Vice versa not possible

**84. How many ports are available in 8051? (L2)**

4

**85. How can a port be configured as an input and output port in 8051? (L2)**

Loading with 1 will set it as input and 0 will be making it as output.

**86. Port – 0 has dual role to play in 8051. What are they? (L2)**

It can act as input or output port and also can help in providing support for address and data when connected to external memory.

**87. What are the 4 modes of timer operations supported in 8051? (L2)**

Mode –0, 1, 2 and 3. (16 bit, 13 bits, 8 bits and split timer)

**88. What is the maximum value that can be loaded in mode 1 of timer operation in 8051? (L2)**

0000H to FFFFH

**89. What is the maximum value that can be loaded in mode 0 of timer operation in 8051? (L2)**

0000H to 1FFFH

**90. What is the maximum value that can be loaded in mode 2 of timer operation 8051? (L2)**

00H to FFH

**91. What is the use of split timer mode in 8051? (L2)**

It becomes two separate 8-bit timers. Precisely, Timer 0 is TL0 and Timer 1 is TH0. Both these timers will count from 0 to 255 and overflow back to 0. All the bits that are related to Timer 1 will now be tied to TH0.

**92. How many timers are available in 8051? (L2)**

2

**93. How will be a particular timer mode selected in 8051? (L2)**

With TMOD register

**94. Timer 1 and Timer 0, both have TH and TL in 8051 – true or false? (L2)**

True

**95. TF in 8051? – What is the usage? (L2)**

Timer flag to indicate the roll over.

**96. How is auto-reload done in 8051? (L2)**

When the count is loaded to the lower byte, it gets loaded to TH as well.

**97. Why is 11.0592 MHz opted in 8051? (L2)**

IBM PC's compatibility is the reason.

**98. How is GATE useful from Timer Mode register in 8051? (L2)**

GATE will help the programmer in selecting software/hardware way of setting and resetting the timer.

**99. How many interrupts are supported in 8051? (L2)**

6

**100. What is an interrupt vector table? (L2)**

It holds the address of ISRs for the interrupts

**101. What is the address for RESET in the Interrupt Vector Table in 8051? (L2)**

0000 H

**102. How can Edge triggered interrupt be set in 8051? (L2)**

Simple, use TCON register and select appropriate bit.

**103. Can IE (Interrupt Enable) register be reset on the whole in 8051? (L2)**

Yes

**104. Which is beneficial serial or parallel communication? (L2)**

Depends. If shorter distance it would be parallel and if longer it will be serial. But, parallel would be expensive.

**105. What are the likely problems in parallel communication? (L2)**

Cost and maintenance are the two major factors.

**106. What RS in RS-232? (L2)**

Recommended Standard.

**107. How is NULL Modem configured? (L3)**

The purpose of a null-modem cable is to permit two RS-232 "DTE" devices to communicate with each other without having any modems or other communication devices (i.e., DCEs) between them. To achieve this, the most obvious connection is that the TD signal of one of the devices must be connected to the RD input of the other device (and vice versa).

**108. How is DB9 different from DB25? (L2)**

Only essential pins out of DB25 are taken and they are framed as DB9.

**109. What is the purpose of signal RTS and CTS in RS 232? (L2)**

RTS – Request To Send – When some data has to be sent from DTE to DCE, RTS will be sent and it will inform DCE that some data is about to be sent and it has to be ready to receive the same. DTE here raises a request to the DCE to be prepared to receive the data. CTS – Clear to send – When the RTS is received, and if the DCE has got enough space to have the data received from DTE to store it. It will send CTS (Clear to send) to make DTE understand that, data transmission can be started. It will serve as an input to DTE from DCE. It says that it is ready to accept the data.

**110. Can RS 232 be used for a long distance communication? (L1)**

No. Can't be used. Because of its physical properties, this is not possible.

**111. What is the problem associated with RS232 and how can it be resolved? (L2)**

It is not TTL compatible. So need MAX 232 chip for solving this issue.

**112. Expand EIA. (L1)**

Electronics Industrial Association.

**113. For setting baud rate, which timer's register has to be used in 8051? (L2)**

Timer 1.

**114. What is the purpose of setting baud rate? (L2)**

The baud rate is the rate used for the communication purpose and it is used to identify that how much data had been transferred and at what speed.

**115. How many addressing modes are there in 8051? (L2)**

8

**116. Define Opcode and operand? (L2)**

Opcode represents the instruction. Operand is part of the instruction. Can be registers or values.

**117. How many operands can be there in an instruction? (L2)**

Depends on the instruction. It can be two or three based on the instruction.

**118. Write an example for immediate addressing mode in 8051. (L2)**

MOV A, #08H

**119. How is direct addressing different from immediate addressing? (L2)**

Immediate addressing will have data as part of the instruction but direct addressing will have the address where the data is stored.

**120. What are the flags affected when arithmetical operations are done? (L2)**

CY, AC and it depends on the operation being done.

**121. Expand DAA used in 8051. (L2)**

Decimal Adjust Accumulator.

**122. Can multiplication be done with any two registers in 8051? (L2)**

No. A and B are the choices.

**123. What are the specific operations which can be done specifically with B register in 8051? (L2)**

Multiplication and division.

**124. MOV A, # 0B0H; If # is not given there in instruction what would have been the impact? (L2)**

It will be treated as address, which will not perform the expected behavior.

**125. After division where the result will be stored in 8051? (L2)**

Quotient will be stored in the Accumulator and remainder in B.

**126. After multiplication what will be stored in B register in 8051? (L2)**

Accumulator. If greater than FF then OV will be set.

**127. What is the instruction used to set the carry in 8051 microcontroller? (L2)**

SETB C.

**128. What is the purpose of CJNE instruction in 8051? (L2)**

Will compare and jump if not equal to the branch.

**129. How are PUSH and POP different? (L2)**

PUSH to add the data to the stack, POP to remove data from the stack.

**130. MOVX, what is that X specifying in 8051? (L2)**

X specifies external memory.

**131. Load 10 to the accumulator and multiply it for ten times, through program. Store the result in R5 with 8051. (L3)**

```
MOV A, #10;
MOV R2,#10;
MUL: ADD A, R1;
DJNZ R2, MUL;
MOV R5, A;
```

**132. Write a program to find 2s complement for a number stored in R1. Move the 2s complemented value to R5 with 8051. (L3)**

```
MOV R1,#10H;
MOV A, R1;
CPL A;
ADD A, #01H;
MOV R5, A;
```

**133. Write a program to swap lower and upper nibble of a number 10H stored in register R1. Move the swapped result to another register R3 with 8051 instruction. (L3)**

```
MOV R1,#10H;
MOV A, R1;
SWAP A;
MOV R3, A;
```

**134. Write a program to exchange content of accumulator with R1's content with 8051 instruction. (L3)**

```
XCH A, R1;
```

**135. Write a program to multiply two 8-bit numbers 08H and 05H. Store the result in R3 with 8051 instruction. (L3)**

```
MOV A, #08H
MOV B, #05H
MUL AB
MOV R3, A
```

**136. Write a program to store the higher nibble of r7 on to both nibbles of r6 with 8051. (L3)**

```
MOVA, R7;        get the content in acc
ANLA, #0F0h;     mask lower bit
MOVR6, A;        send it to r6
SWAPA;           Exchange upper and lower nibbles of acc
ORLA, R6;        logical OR operation
MOVR6, A;        finally load content in r6
```

**137. Write a program with 8051 instruction that could divide the content of r1 by r2. Then result in r3 and r4 with quotient and remainder. Then restore the original content of r1. (L3)**

```
MOV A, R1;       Move the content of R1 and R2 to A and B
MOV B, R2;
```

```
DIV AB;        divide A by B
MOV R3, A;     store quotient in R3
MOV R4, B;     reminder in R4
MOV B, R2;     again get content of R2 in B
MUL AB;        multiply it by quotient
ADD A, R4;     add reminder in new answer
MOV R1, A;     restore the content of R1
```

**138. What is step angle? (L1)**

Step angle is defined as an angle through which motor shaft rotates in one single step.

**139. What is the unit in which stepper motor's speed is measured? (L1)**

RPM (rotations per min)

**140. How is stepper motor different from DC motor? (L1)**

Stepper motor rotates in fixed steps whereas DC motor moves continuously.

**141. How can one make a stepper motor rotate in anticlockwise direction? (L1)**

Loading the sequence in revere direction will get this task accomplished.

**142. What is the basic principle used behind interfacing the DC motor with 8051? (L1)**

H – Bridge.

**143. Expand ARM. (L1)**

ACRON RISC MACHINE.

**144. Which one is preferred? RISC or CISC? (L2)**

Cannot conclude. They both are good and still they exist in the market. Based on the requirement one would go for CISC or RISC.

**145. What is the architecture followed in ARM for processing the instructions? (L2)**

Load and Store.

**146. What is the basic architecture followed by ARM7 and ARM9? (L1)**

Von-Neumann and Harward.

**147. Which registers in the ARM register is referred as Program Counter? (L3)**

R15

**148. What are the different states that an ARM core can work on? (L3)**

ARM, THUMB and JAZELLE

**149. What is the purpose of CPSR in ARM? (L3)**

CPSR will be used to set the mode of operations also it is used to check the status of the operations carried out.

**150. What are the flags supported in CPSR in ARM? (L3)**

Negative, Carry, Overflow and Zero flags.

**151. How will one set the THUMB state in ARM? (L3)**

Setting the THUMB bit T from CPSR will get the state set to THUMB mode.

**152. How can JAZELLE state be set in ARM?**

Jazelle supported cores will have J bit in the CPSR. Setting it will get the Jazelle mode enabled.

**153. What is SPSR? Is it supported for all the modes in ARM?**

SPSR is expanded as Saved Program Status Register. It is specially meant for privileged modes of operation. There is an exception here; system mode does not have support for SPSR

**154. What are the different modes of operation supported in ARM? (L2)**

- Abort mode
- Fast interrupt request mode
- Interrupt request
- Supervisor mode
- System mode
- Undefined mode
- User mode

**155. What is a privileged mode in ARM? What are the modes that fall in this category? (L2)**

Privileged modes will have better access permissions. Following are the privileged modes.

- Abort mode
- Fast interrupt request mode
- Interrupt request
- Supervisor mode
- System mode
- Undefined mode

**156. What are the different pipelining modes followed in ARM? (L2)**

3, 5 and 6 stage pipelining.

**157. Write in brief on the pipelining in ARM? (L3)**

Fetch: Fetching the instruction for execution from memory, where there will be pool of instructions, which is nothing but the program or the code written by the programmer.

Decode: Instruction which is fetched needs to identified on what it is? What is meant for? What kind of operation would it perform? etc., this process is referred to be as decoding and the decode stage does this work.

Execute : Result will be generated out of this step.

Buffer/memory: ALU result will be buffered for a clock cycle here which can facilitate pipeline flow for all the instructions.

Write: Results obtained due to the execution are moved to the registers.

**158. What are the exceptions supported in ARM? (L3)**

Following are the exceptions supported in ARM

- Reset
- Undefined instruction
- Software interrupt
- Prefetch abort
- Data abort
- Reserved
- Interrupt request

**159. Pick odd one out! (Embedded is the clue) (L1)**

(a) **Laptop**
(b) projector
(c) Mobile phone
(d) MP3 player

**160. Some of the important characteristics expected in consumer electronics products are ____________ (L1)**

(a) Recovering from failures
(b) Low cost
(c) Performance
(d) **Low unit cost, low power consumption and smaller product size**

**161. One of the most important characteristics of embedded systems for automotive industry (with respect to the inmates of the vehicle) is ____________ (L1)**

(a) Recovering from failures

(b) Low cost

(c) Safety

**162. Give an example of multi-rate characteristic of an Embedded System! (L2)**

A digital camera can work on a black and white image, color image, video and audio. Processing is different in all these which is referred as multi-rate.

**163. Embedded systems are almost battery operated – True or False. (L1)**

True.

**164. Mobile phone is not an Embedded System – True or False. (L1)**

False.

**165. Hardware is tangible, but software is intangible – right or wrong? (L1)**

Yes.

**166. How is it possible that both the programs and data can be stored on the same floppy disk? (L1)**

(a) A floppy disk has two sides, one for data and one for programs.

(b) Programs and data are both software, and both can be stored on any memory device.

(c) A floppy disk has to be formatted for one or for the other.

**167. What are the two general types of programs? (L1)**

(a) Entertainment and Productivity.

(b) Microsoft and IBM.

(c) System software and Application software.

**168. Why a processor does not contain a cache of huge size? (L1)**

(a) Costly

(b) Spacious

(c) Not available in the market

(d) No use

**169. What is tool used for debugging the software on a hardware? (L1)**

(a) JTAG

(b) BTAG

**170. For designing an embedded system which of the following modeling is preferred (L1)**

(a) Waterfall model

(b) Spiral model

(c) RAD

(d) Consecutive refinement model

**171. The most important phase in software life cycle is (L1)**

(a) Integration
(b) **Design**
(c) Testing
(d) None of the above.

**172. A very important advantage associated with simulation is( L1)**

(a) **Improved safety and reliability**
(b) Reduced cost factor
(c) Reduced defects
(d) Reduced setting up time

**173. A major problem with simulation is (L1)**

(a) Simulators are sometimes expensive
(b) Setting up a simulator may take time
(c) **Cannot test with all possible inputs**
(d) Though simulated, needs a physical system to test completely.

**174. Which of the following is an RTOS? (L1)**

(a) **VxWorks**
(b) Linux
(c) Ubuntu
(d) Windows XP

**175. Which of the following is most important and expected behavior of an RTOS? (L1)**

(a) Reduced memory usage
(b) Multitasking
(c) **Response time**
(d) Protection from the applications.

**176. Which of the following can't be installed in an Embedded System? (L1)**

(a) **Windows XP**
(b) VxWorks
(c) WinCE

**177. Cache memory is faster and costlier than RAM- true/false. (L1)**

True.

**178. Differentiate preemptive and non-preemptive techniques. (L1)**

Preemptive takes into account the priority of the processes, non-preemptive doesn't.

**179. Differentiate binary semaphore and counting semaphore. (L1)**

Binary semaphore just has two values 0/1 but counting semaphore deals with a large set of values.

**180. Protocols may include which of the following (L1)**

(a) Signaling
(b) Authentication
(c) Error-checking
(d) **All of these**

**181. A device driver simplifies programming by acting as translator between ____________ or operating systems that use it (L1)**

(a) <u>**Hardware device and the applications**</u>
(b) Software and applications
(c) None of these

**182. Virtual serial port emulation can be useful in case there is a lack of available ____________ ports or they do not meet the current requirements. (L1)**

(a) <u>**Physical serial**</u>
(b) Data
(c) Network
(d) None of these

**183. Some synchronous devices provide a ____________ to synchronize data transmission, especially at higher data rates. (L1)**

(a) Data Signal
(b) Timer
(c) <u>**Clock signal**</u>
(d) No signal

# 2 DIGITAL ELECTRONICS

These are the few basic interview questions from the area of digital electronics. After reading electronics, it becomes mandatory for anyone to know these basics.

**1. Explain the need for learning digital electronics? (L2)**

One may read a book in Kindle as an e-book or as a PDF in smart phone. The smart phone, the laptop, the air-conditioning systems are all the gifts of digital electronics. The applications of digital electronics are widespread. Take a simple pocket calculator to the high end rocket systems; they are all outcomes of digital electronics. So, it is highly important to learn and understand digital electronics to understand the world better.

**2. What is De-Morgan's law? Enumerate. (L2)**

De-Morgan specified the following two laws which serve as the base for digital electronics.

$$\overline{AB} = \overline{A} + \overline{B}$$

$$\overline{A + B} = \overline{A} \cdot \overline{B}$$

**3. Give a note on the following number systems. (L3)**

Decimal, Binary, Octal, Hexadecimal, BCD, ASCII, EBCDIC, Gray code.

- ✓ A decimal number can be any number ranging from 0 to infinity without fractional parts and are represented by the number to the base 10. Example: $(45)_{10}$.
- ✓ A binary number is composed of only 1s and 0s and are represented by the number to the base 2. Example: $(1011\ 0100)_2$.
- ✓ An octal number can be any number from 0 to infinity without fractional parts and are represented in terms of the number to the base 8. Example: $(67)_8$.
- ✓ A hexadecimal number is represented with the number to the base 16. Example: $(AB3)_{16}$.
- ✓ BCD stands for Binary Coded Decimal. Any number from 0 to 9 can be referred to as a BCD number. A number greater than 9 can be converted into BCD format by adding 6 to the number.
- ✓ ASCII stands for American Standard Code for Information Interchange. For every letter in the alphabet (upper or lower case), for every numeric character,

for every special symbol, there is an ASCII equivalent decimal. For example, ASCII equivalent for A is $65_{10}$ and for a is $97_{10}$.

- ✓ EBCDIC stands for Extended Binary Coded Decimal Interchange Code. This is the code predominantly used in IBM mainframe computers.
- ✓ Gray code is a binary numeral system where two successive values differ in only one bit.

**4. How is a positive number represented in Binary format? (L2)**

A positive number can be represented using binary number system using 8 4 2 1 weighted binary format. For an instance, if 14 has to be converted to binary value, it can be achieved with following steps:

| 8 | 4 | 2 | 1 |
|---|---|---|---|
| 1 | 1 | 1 | 0 = 1110 |

8 + 4 + 2 + 0 = 14.

$(14)_{10} = (1110)_2$

**5. How is the negative number represented in Binary format? (L2)**

A negative number can be represented using binary number system using the following steps:

- Convert the corresponding positive number into binary number using 8 4 2 1 weighted binary format.
- Take one's complement of the obtained binary number. One's complement is obtained by reversing all 1s to 0s and 0s to 1s.
- Take two's complement of the obtained one's complement number by adding 1 to the one's complement number.

  An example to convert – $(73)_{10}$ to binary.
- Now to get the binary representation of – $(73)_{10}$, add 1 to the found one's complement result.
- 73 is represented in binary as follows:

| 128 | 64 | 32 | 16 | 8 | 4 | 2 | 1 |
|---|---|---|---|---|---|---|---|
| 0 | 1 | 0 | 0 | 1 | 0 | 0 | 1 = 64 + 8 + 1 = 73. |

```
0 1 0 0 1 0 0 1   <----  binary representation

1 0 1 1 0 1 1 0   <----  one's complement
                +
              1          (0 + 1 = 1) refer to section 6.3.1
_______________

1 0 1 1 0 1 1 1   <----  two's complement
_______________
```

**6. How can the 4-bit binary sequence be referred as?**

The 8 4 2 1 code.

7. **What is the decimal equivalent of the binary number 1011?**

   11.

8. **In a normal binary sequence, which bit is more prone to variations in the sequence than the other bits?**

   The Least Significant Bit (LSB).

9. **Which bit in a binary sequence varies less frequently compared to the other bits in the sequence?**

   The Most Significant Bit (MSB).

10. **What is a digital system comprised of? (L1)**

    Mostly Logic gates. Other than that there would be other components as resistors, capacitors and connecting wires would be part of any circuit.

11. **Which is the conventional logic system? (L1)**

    Positive logic system (HIGH = 1; LOW = 0). This is the mostly followed convention across all the digital systems.

12. **Which simplified representation of a logic function depicts all the input – output relations of a logic function? (L1)**

    Truth table.

13. **What are the values associated with Boolean algebra? (L1)**

    Boolean algebra consists of 0's and 1's only.

14. **What is TTL? State the voltage ranges of different TTL states. (L1)**

    TTL stands for Transistor Transistor Logic. According to the TTL standards, voltage of LOGIC HIGH state ranges from 2 V to 5 V and the voltage range of LOGIC LOW state is between 0 V and 0.8 V.

15. **What does CMOS stand for? Give the voltage ranges of the different states defined by the CMOS logic? (L1)**

    CMOS stands for Complementary Metal Oxide Semiconductor. In CMOS logic, voltage of HIGH state ranges from 3.5 V to 5 V while the voltage range of LOW state is 0 V to 1.5 V.

16. **Name the three basic logic operations? (L1)**

    NOT, AND and OR are the basic logic operations which lead to many further operations.

17. **What is a logic gate? Draw its symbol. (L2)**

    Logic gate is a basic building block of a digital system. A gate has one or more inputs and only one output. A logic gate processes the input logic levels HIGH (1) or LOW (0) and produces a logic output (1 or 0) depending upon the logic function definition of the gate. Figure 2.1 represents the symbol of Logic Gate.

Fig. 2.1 Logic gate – Block diagram

**18. Which logic operation results in an output which gives the complement of the input? (L1)**

NOT operation or inversion. We call it invertor. Based on the truth table shown below one can understand this.

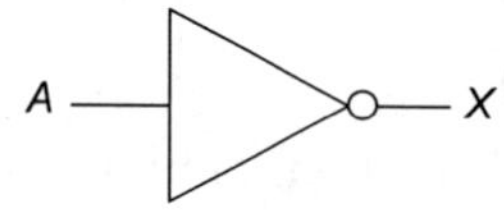


(a) Logic symbol

| Input *A* | Output *X* |
|---|---|
| 0 | 1 |
| 1 | 0 |

(b) Truth table

**19. Draw the electrical circuit equivalent for NOT gate. (L3)**

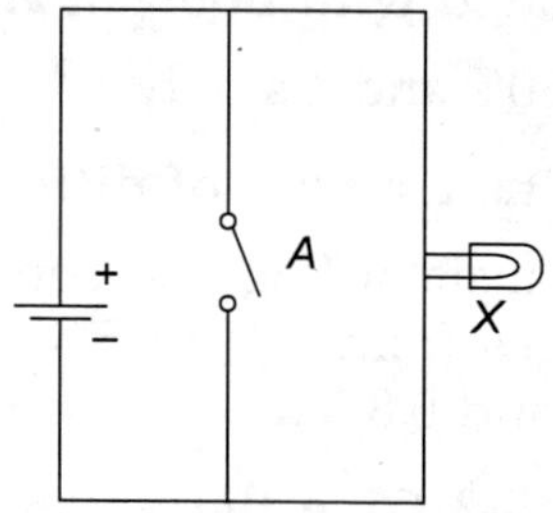


Fig. 2.2 Circuit equivalent of NOT gate.

One can understand the way the above circuit works easily. When the SWITCH is OPEN ($A = 0$), the voltage at the battery appears across the LED and the LED GLOWS ($X = 1$). When the SWITCH is CLOSED ($A = 1$), the voltage across the battery is shorted through the SWITCH and no current flows through the LED and hence it remains OFF ($X = 0$). Figure 2.2 represents the circuit equivalent.

**20. Which logic operation produces a HIGH output only when all the inputs are HIGH? (L3)**

AND operation.One can understand the way the AND gate works through the truth table revealed below.

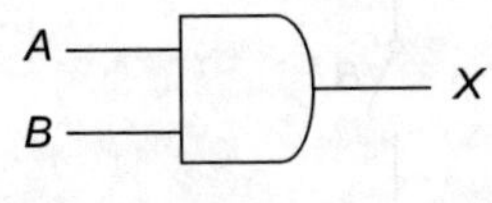


(a) Logic symbol

| Inputs | | Output |
|---|---|---|
| A | B | X |
| 0 | 0 | 0 |
| 0 | 1 | 0 |
| 1 | 0 | 0 |
| 1 | 1 | 1 |

(b) Truth table

**21. Draw the electric circuit equivalent for AND gate. (L2)**

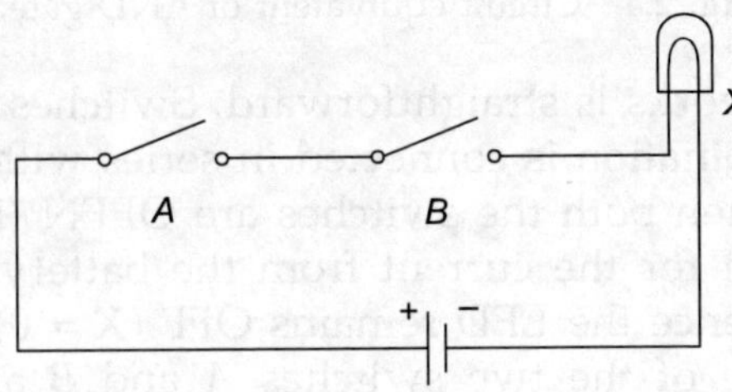


**Fig. 2.3** Circuit equivalent of AND gate.

It is clearly visible that, when both *A* and *B* are OPEN (i.e., *A* = 0, *B* = 0), there is no path for the current to flow. The circuit is open and the LED remains OFF (*X* = 0). Even when any one of the switches are CLOSED (i.e., (*A*, *B*) = (0, 1) or (1, 0)), the same condition prevails and the LED remains OFF (*X* = 0). Now, if both the switches are CLOSED (i.e., *A* = 1 and *B* = 1), the circuit becomes closed. The current flows through the LED and it GLOWS (*X* = 1).

**22. Which is the AND gate IC? (L3)**

IC 7408.

**23. When there is no enough light on the streets, which logic gate can be used to reverse the condition based on automatic street lighting system? (L2)**

NOT gate.

**24. Which is the NOT gate IC? (L2)**

IC 7404.

**25. Which logic operation produces a HIGH output when at least one of the inputs is HIGH? (L2)**

OR gate.

**26. Draw the circuit equivalent of OR Gate. (L3)**

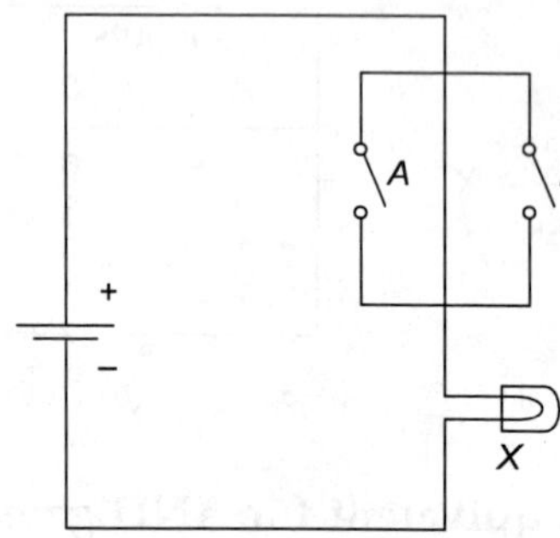


**Fig. 2.4** Circuit equivalent of AND gate.

The way the circuit works is straightforward. Switches *A* and *B* are connected in parallel and the combination is connected in series with the LED and the battery. Consider the case when both the switches are OPEN (i.e. (*A*, *B*) = (0, 0)). In this case, there is no path for the current from the battery to flow through and the circuit is open and hence the LED remains OFF (*X* = 0). Now let us examine the case when either one of the two switches *A* and *B* are CLOSED (i.e. (*A*, *B*) = {(0, 1) or (1, 0)}). When any of the switches are CLOSED, the circuit gets complete and there is a path for current flow and hence, the LED GLOWS (*X* = 1). Similarly, when both the switches are CLOSED (i.e. *A* = *B* = 1), there exists a closed path for the current flow and the LED GLOWS (*X* = 1). Figure 2.4 reveals the circuit equivalent.

**27. Which is the OR gate IC? (L2)**

IC 7432.

**28. Which gate is the transparent gate? (L2)**

The buffer gate. It just produces delay. Does not do much with data bits.

**29. What is the main purpose of the buffer gate? (L2)**

To produce the required time delay one would use the buffer gate.

**30. Draw the circuit equivalent of BUFFER gate. (L2)**

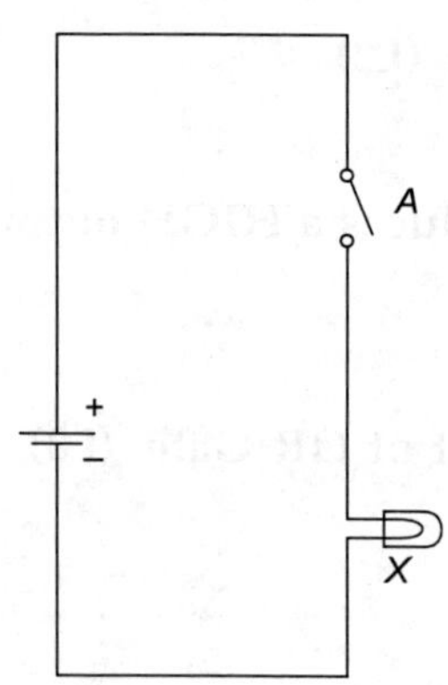


**Fig. 2.5** Circuit equivalent of Buffer gate.

This is one of the simplest circuits that one would think of. When the switch is OPEN ($A = 0$), the LED remains OFF (X = 0). When the switch is CLOSED ($A = 1$), the circuit gets closed and the LED GLOWS (X = 1). Figure 2.5 represents the same. To make it simple, when you turn the switch on, the light will glow.

**31. Which are the universal gates. (L2)**

NAND and NOR.

**32. Which is the NAND gate IC? (L2)**

IC 7400.

**33. Explain the way NAND gate functions. (L2)**

AND with Inverter is referred as NAND gate. When any one of the inputs are true, the output will remain HIGH. When both the inputs are HIGH, the output will remain LOW. The truth table is presented for reference.

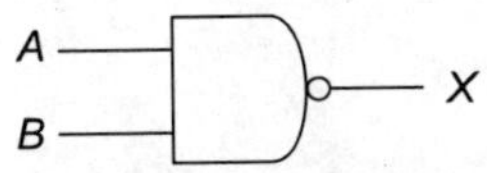


(a) Logic symbol

| Inputs | | Output |
|---|---|---|
| *A* | *B* | *X* |
| 0 | 0 | 1 |
| 0 | 1 | 1 |
| 1 | 0 | 1 |
| 1 | 1 | 0 |

(b) Truth table

**34. Draw the circuit equivalent of NAND gate. (L2)**

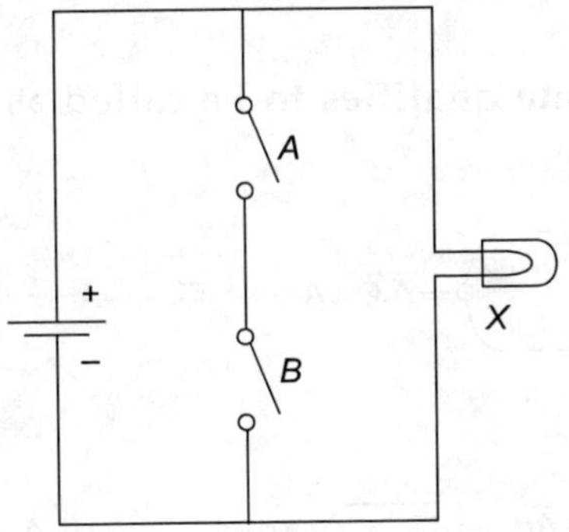


**Fig. 2.6** Circuit equivalent of a NAND gate.

The explanation is slightly complex when compared to the previous set of explanations. When both *A* and *B* remain open, the current flow is through the branch containing LED and it would glow. If any of the switches *A* or *B* is open, the same case as previous prevails. When both the switches *A* and *B* are CLOSED ($A = B = 1$), there is a shorted path for the current from the battery to flow through and hence there is no current flow through the LED. Thus, LED doesn't glow (X = 0).

**35. How NAND can be constructed with AND gate? (L2)**

It is simple. One can have an inverter connected to an AND gate to get it converted as a NAND gate. The bubble drawn over the AND gate makes it NAND. Figure 2.7 represents the same.

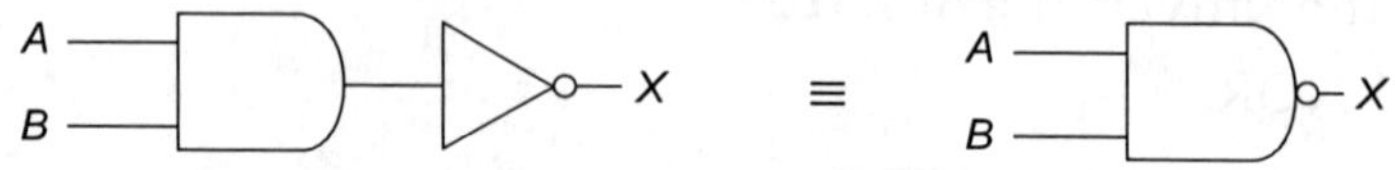


**Fig. 2.7** AND to NAND conversion.

**36. Which is the NOR gate IC? (L2)**

IC 7402.

**37. Which is the XOR gate IC? (L2)**

IC 7486.

**38. Which is the XNOR gate IC? (L2)**

IC 747266.

**39. What makes up a CMOS? (L2)**

Combination of a P-MOS and an N-MOS.

**40. Which type of MOSFET offers voltage pull-up? (L2)**

P-MOS.

**41. Which type of MOSFET is suitable for voltage pull-down? (L2)**

N-MOS.

**42. Justify how NAND Gate qualifies to be called as Universal Gate? (L3)**

Case: 1 NAND as NOT.

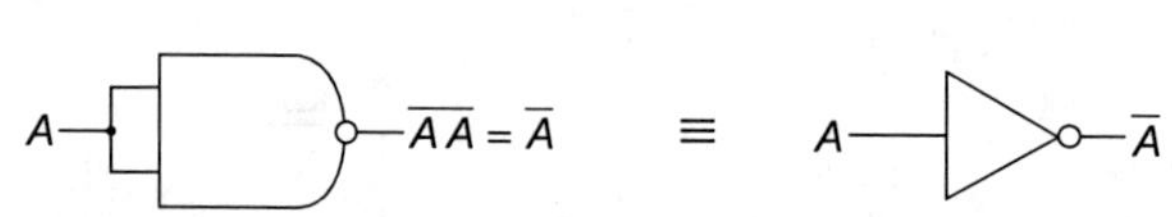


Case: 2 NAND as AND.

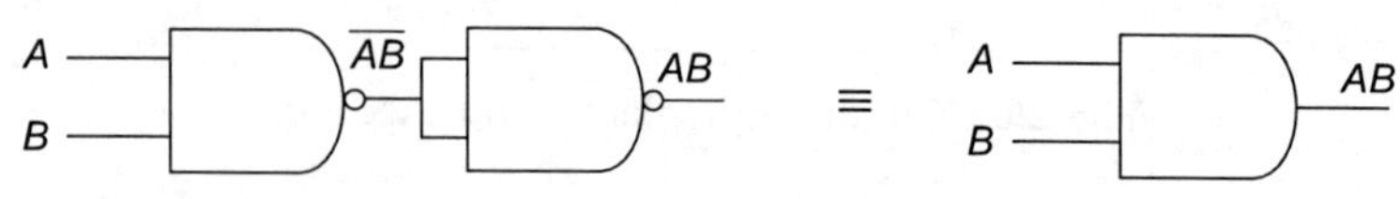


Case: 3 NAND as OR.

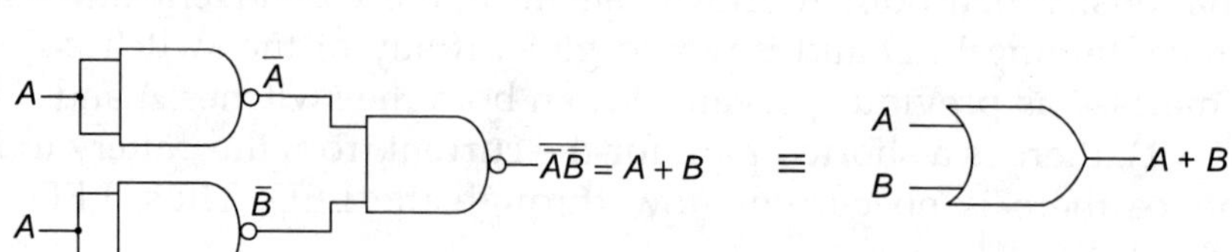

Above three cases proves that NAND gate can be useful in making NOT, AND and OR gates. So it is termed Universal Gate.

**43. In which logic system does a relatively negative voltage state correspond to a HIGH state and the vice versa? (L2)**

Negative logic system.

**44. List the key differences between combinational and sequential circuits. (L3)**

| S. No | Combinational Circuit | Sequential Circuit |
|---|---|---|
| 1 | There is no memory element in the combinational circuit. | Memory elements are one of the core components of the sequential circuits. |
| 2 | The current state is dependent only on the present state inputs. | The current state is dependent on the previous state and also the current state. |

**45. What does the output of a combinational circuit depend on? (L2)**

Present input only.

**46. What does the output of a sequential circuit depend on? (L2)**

Present and past inputs.

**47. What is the most important component of a sequential circuit? (L2)**

Memory, without which the circuit is not going to work in the way it is expected to work.

**48. What is an adder? (L2)**

Adder is used to perform arithmetic addition operation of bits and the obtained Sum and Carry are given as outputs.

**49. What is the difference between a half adder and a full adder? (L2)**

Half adder does not support carry whereas full adder supports it.

**50. Draw the truth table for half-adder which would help in deriving the final expression and also the circuit. (L2)**

| Input B | Input A | Carry digit | Sum digit |
|---|---|---|---|
| 0 | 0 | 0 | 0 |
| 0 | 1 | 0 | 1 |
| 1 | 0 | 0 | 1 |
| 1 | 1 | 1 | 0 |

**51. What is final expression for Sum and Carry in half adder, also draw the logic circuit?**

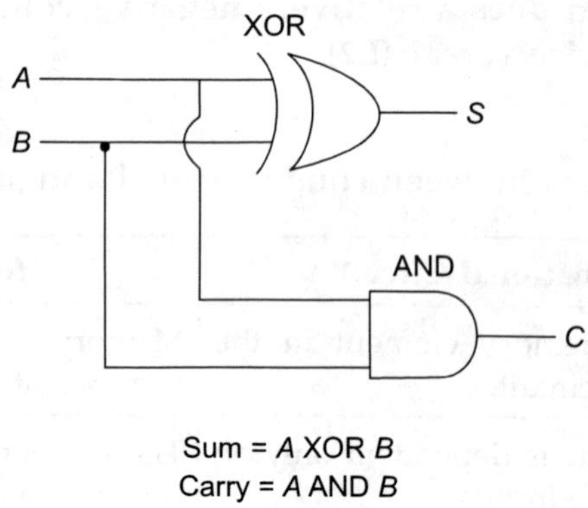


**Fig. 2.8** Circuit for half adder.

Final expressions for SUM and CARRY respectively for the half adder is

Sum = $A$ XOR $B$ whereas Carry = $A$ AND $B$.

**52. Draft the truth table for the full adder. (L3)**

| Carry in | Input *B* | Input *A* | Carry out | Sum digit |
|---|---|---|---|---|
| 0 | 0 | 0 | 0 | 0 |
| 0 | 0 | 1 | 0 | 1 |
| 0 | 1 | 0 | 0 | 1 |
| 0 | 1 | 1 | 1 | 0 |
| 1 | 0 | 0 | 0 | 1 |
| 1 | 0 | 1 | 1 | 0 |
| 1 | 1 | 0 | 1 | 0 |
| 1 | 1 | 1 | 1 | 1 |

**53. What is the final expression for sum and carry in the full adder design? (L2)**

For Sum:

$$S = A \oplus (B \oplus C_{in})$$
$$S = A \oplus B \oplus C_{in}$$

For Carry:

$$\text{Carry} = \bar{A}BC_{in} + A\bar{B}C_{in} + AB$$

or

$$= C_{in}(A \oplus B) + AB \quad (1)$$

**54. Draw the logic circuit for design of a full adder. (L3)**

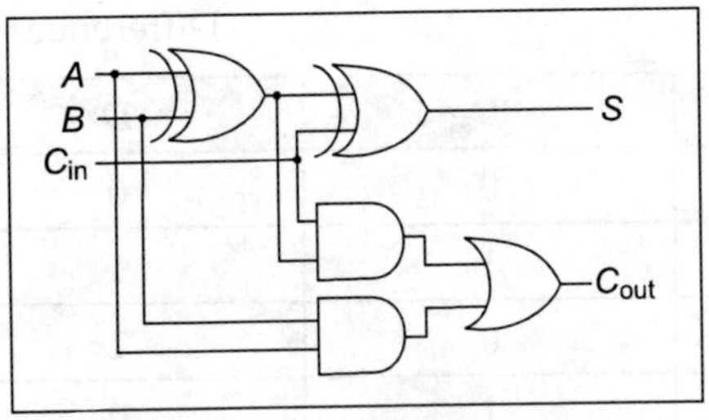


**Fig. 2.9** Circuit for full adder.

Keeping the expressions for Sum and Carry as reference the above logic circuit can be drawn as shown in Fig. 2.9.

**55. Draw the symbol of 4-bit ripple carry adder. (L2)**

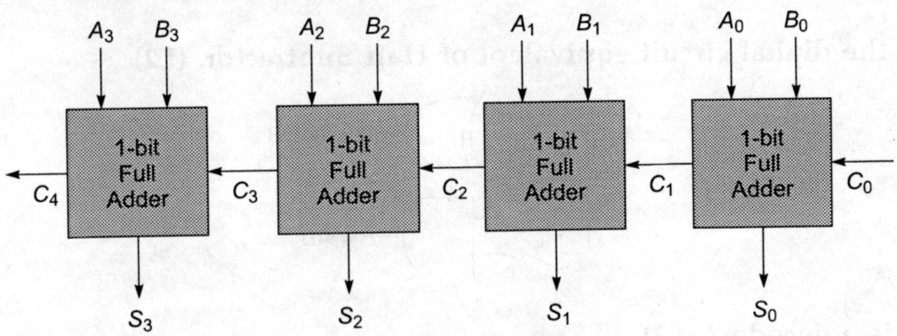


**56. Draw the digital circuit equivalent for the 4-bit ripple carry adder. (L3)**

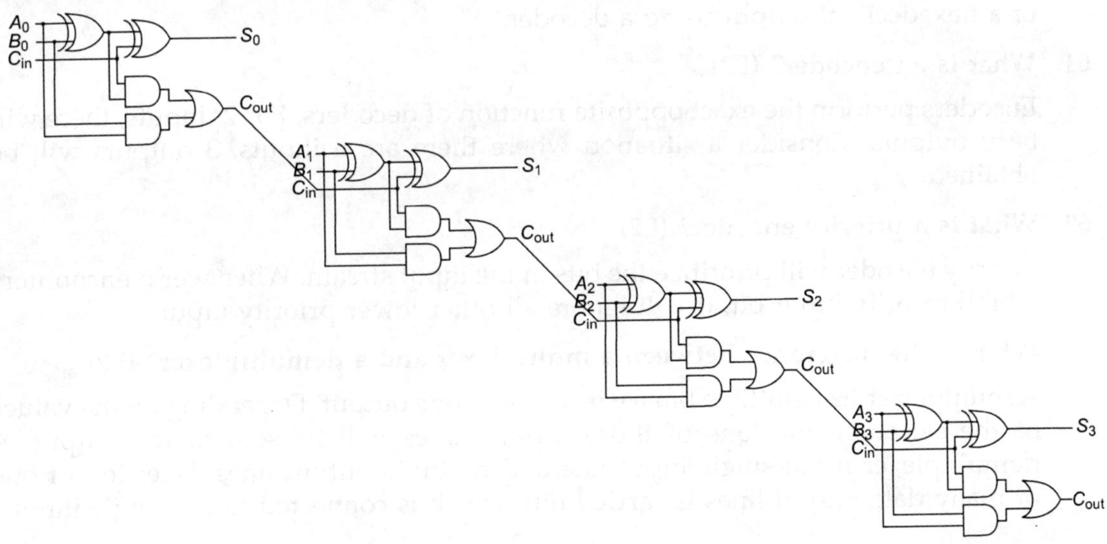

**57. Draft the truth table of Half Subtractor. (L2)**

| | | Difference | Borrow |
|---|---|---|---|
| *A* | *B* | *D* | *B* |
| 0 | 0 | 0 | 0 |
| 0 | 1 | 1 | 1 |
| 1 | 0 | 1 | 0 |
| 1 | 1 | 0 | 0 |

**58. What is the final expression for difference and borrow in the half subtractor design? (L2)**

$$D = \bar{A}B + A\bar{B} = A \oplus B$$
$$B = \bar{A}B$$

**59. Draw the digital circuit equivalent of Half Subtractor. (L2)**

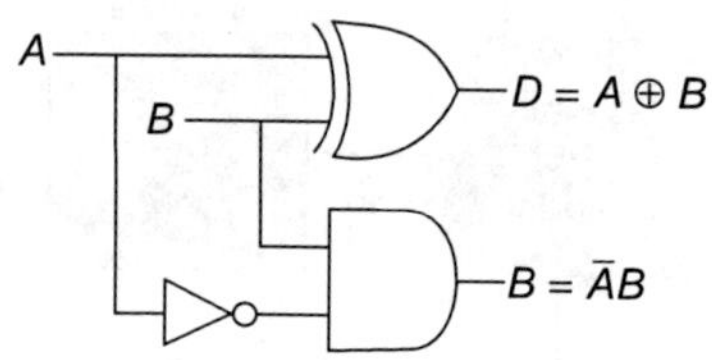


**60. What is a decoder? (L2)**

When $n$-bit binary information is fed into a Decoder, it converts it to $2^n$ unique outputs. In short, a binary code can be converted either to a decimal or an octal or a hexadecimal output using a decoder.

**61. What is an encoder? (L2)**

Encoders perform the exact opposite function of decoders. For $2^n$ inputs, there will be $n$ outputs. Consider a situation where there are 8 inputs, 3 outputs will be obtained.

**62. What is a priority encoder? (L2)**

Priority encoder will prioritize the bits in the input stream. Whenever it encounters a high priority input bit, it will ignore all other lower priority inputs.

**63. What is the difference between a multiplexer and a demultiplexer? (L2)**

A multiplexer has multiple binary inputs and one output. Depending on the values of the selection line, one of those input values will be sent to the output. A demultiplexer has a single input line and multiple output lines. Selection of one of many data-output-lines is carried out, which is connected to the single input.

**64. Draw the truth table of 3 to 8 decoder. (L3)**

| $A_2$ | $A_1$ | $A_0$ | $O_7$ | $O_6$ | $O_5$ | $O_4$ | $O_3$ | $O_2$ | $O_1$ | $O_0$ |
|---|---|---|---|---|---|---|---|---|---|---|
| 0 | 0 | 0 | 0 | 0 | 0 | 0 | 0 | 0 | 0 | ① |
| 0 | 0 | 1 | 0 | 0 | 0 | 0 | 0 | 0 | ① | 0 |
| 0 | 1 | 0 | 0 | 0 | 0 | 0 | 0 | ① | 0 | 0 |
| 0 | 1 | 1 | 0 | 0 | 0 | 0 | ① | 0 | 0 | 0 |
| 1 | 0 | 0 | 0 | 0 | 0 | ① | 0 | 0 | 0 | 0 |
| 1 | 0 | 1 | 0 | 0 | ① | 0 | 0 | 0 | 0 | 0 |
| 1 | 1 | 0 | 0 | ① | 0 | 0 | 0 | 0 | 0 | 0 |
| 1 | 1 | 1 | ① | 0 | 0 | 0 | 0 | 0 | 0 | 0 |

**65. What is the purpose of having enable line in any decoder? (Enable Signal) (L2)**

When Enable is 0, selected output will be active. Else, no output would not be. Rest of the functionalities remain the same with a decoder having no Enable signal.

**66. Draw the truth table of 4 to 2 Encoder. (L2)**

| $A_3$ | $A_2$ | $A_1$ | $A_0$ | $F_1$ | $F_0$ |
|---|---|---|---|---|---|
| 0 | 0 | 0 | ① | 0 | 0 |
| 0 | 0 | ① | 0 | 0 | ① |
| 0 | ① | 0 | 0 | ① | 0 |
| ① | 0 | 1 | 1 | ① | ① |

$A$ = Input

$F$ = Output

**67. Name a flip-flop which would be best suited in designing counters. (L2)**

The Toggle flip-flop (T-flip-flop). A T-flip-flop can be implemented by shorting the input terminals of a JK flip-flop.

**68. Clock pulse triggering is used to classify flip-flops. How can they be classified? (L2)**

Positive edge triggered and negative edge triggered. (Level triggering is also one of the classification but it has not been dealt in the counters design.)

**69. Which type of flip-flops produces outputs only when there is a clock transition from LOW to HIGH (a rising edge)? (L2)**

The positive edge triggered flip-flops.

**70. Which type of flip-flops produces outputs only when there is a clock transition from HIGH to LOW (a falling edge)? (L2)**

The negative edge triggered flip-flops.

**71. Which type of counters use a common clock signal at all its flip-flops? (L2)**

Synchronous counters.

**72. How is a counter named when the flip-flops in the counter do not use a common clock signal? (L2)**

Asynchronous counter.

**73. Name the counter which counts ten states repeatedly from 0 0 0 0 to 1 0 0 1? (L2)**

A decade counter.

**74. Ripple counter which is based on T – flip-flop is part of which type of counters? (L2)**

Asynchronous counters.

**75. Undefined states attributed due to propagation delays are possible in which type of counters? (L2)**

Asynchronous counters.

**76. Ring counter falls under which type of counters? (L2)**

Synchronous counters.

**77. Johnson counters are synchronous counters. (True/False) (L2)**

True.

**78. How is the number of unique states that a counter output takes before the sequence repeats itself termed as? (L2)**

The counter modulus or the modulus of the counter.

**79. What are the states that a Modulus 5 counter would take? (L2)**

0, 1, 2, 3 and 4.

**80. State the name of the counter modulus, if a counter takes the states 0, 1, 2, 3, 4, 5 and 6 repeatedly. (L2)**

The counter modulus is 7 (i.e., it is a modulus 7 counter.)

**81. In the SISO shift register, which flip-flop can be used as registers? (L2)**

D-flip-flop.

**82. When does the shifting happen? (L2)**

During the clock pulse the bits will be shifted.

**83. Briefly describe how the SIPO Shift register works? (L2)**

The data and the shifting operation of SIPO is similar to that of SISO shift register. The data is fed bit by bit. While shifting, the only difference is that once all the four bits are shifted, the output is obtained at one shot, in parallel, simultaneously. This is referred as SIPO shift register.

**84. Define Flip-Flop. (L2)**

A flip-flop is a single bit storage element which can store one bit of information when given power supply. As long as power is fed in, the storage will remain. Else, the data is lost.

**85. How is flip-flop different from latch? (L2)**

Flip-flops are different from latch. The output of the latch is affected consistently by the inputs when the enable is asserted. But, when we take flip-flop into consideration, the output will be changed during rising or falling edge of the clock pulse.

**86. What are the types of flip flops you are aware of? (L1)**

D-Flip flop, JK-Flip-flop, T-Flip-flop, SR Flip-flop are the types of flip-flop.

**87. Sketch the symbol and truth table of D-Flip-flop. (L2)**

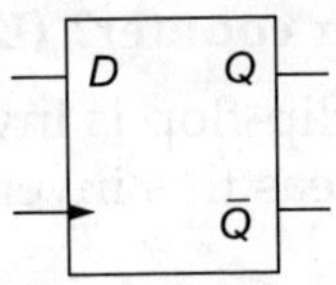


| Clock | D | Q |
|---|---|---|
| Rising edge | 0 | 0 |
| Rising edge | 1 | 1 |

**88. Define setup time and hold time. (L2)**

**Setup time** is defined as the minimum amount of time the data signal should be held steady before the clock event so that the data are reliably sampled by the clock.

**Hold time** is the minimum amount of time the data signal should be held steady after the clock event so that the data are reliably sampled.

**89. Check the diagram below and determine if it is a valid flip-flop?**

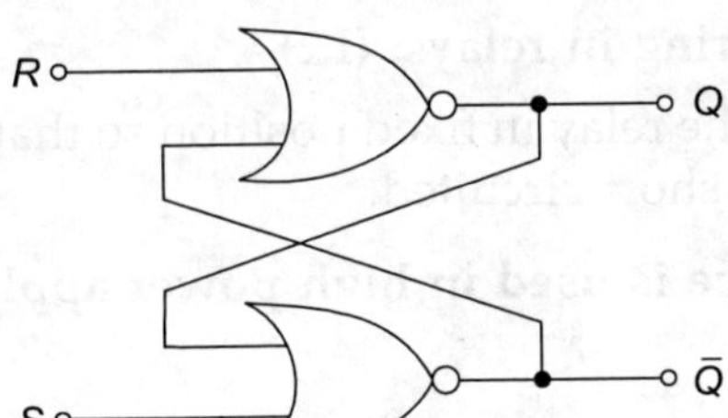


Yes. It is a valid representation. The symbol denotes RS flip-flop.

**90. In JK flip-flop, what could be the output when J and K both remain 0? (L2)**

The output stays the same as it was before. Here the clock pulse will not be effective at all.

**91. Draw symbol of JK Flip-flop. (L2)**

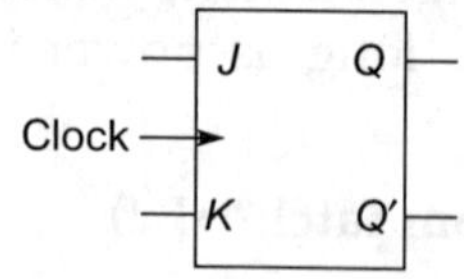


**JK Flip-flop**

**92. How did ring counter get its name? (L2)**

It may be visualized as a circulating shift register. The output of the MSB (last flip flop in the circuit) will be fed back to the input of the first flip-flop, i.e., LSB. A 4-bit ring counter would be sufficient to understand the concept. The circuit is designed in such a way that only one is kept circulated and it is done in synchronization with the clock pulses.

**93. How is a ring counter different from Johnson counter? (L2)**

In a Johnson counter, the output from the last flip-flop is inverted and fed into the first one, and then shifting is carried out, whereas this inversion is not carried out in case of normal ring counter.

**94. Which IC can perform the shifting operations? (L2)**

74LS95 by Motorola

**95. What is the speciality of a universal shift register? (L2)**

It can serve as a shift right register, shift left register and parallel input parallel output register and hence it is called universal.

**96. Relays cannot operate on which kind of faults? (L2)**

External or through faults.

**97. Describe the use of spring in relays. (L2)**

To hold the contacts of the relay in fixed position so that when the coil de-energizes, the contacts do not get short circuited.

**98. Which protection device is used in high power application? (L2)**

Contactors

**99. Why is reed relay preferred over the conventional ones? (L2)**

They switch faster because the contacts are placed in an inert gas filled tube.

**100. What is the need for time delay relays? (L2)**

To make relays work with time delay during closure or opening as per the requirement.

**101. Where is protective relays commonly used? (L2)**

Mostly in industries where heavy power is dealt with, protective relays are deployed.

**102. How can solid state relays be distinguished from the conventional ones? (L2)**

Solid state relays ensure faster switching due to usage of semiconductor devices with no movable parts.

**103. Draw the symbol of solid state relay. (L2)**

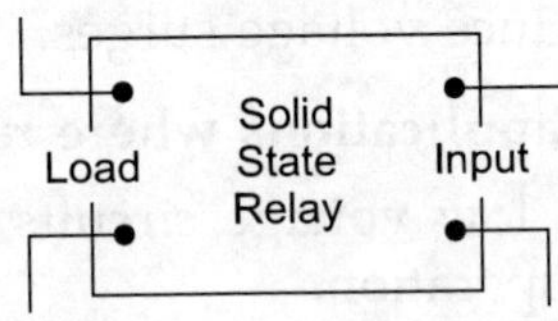


**104. Draw the timing diagram for normally closed-timed closed (NCTC) relay. (L2)**

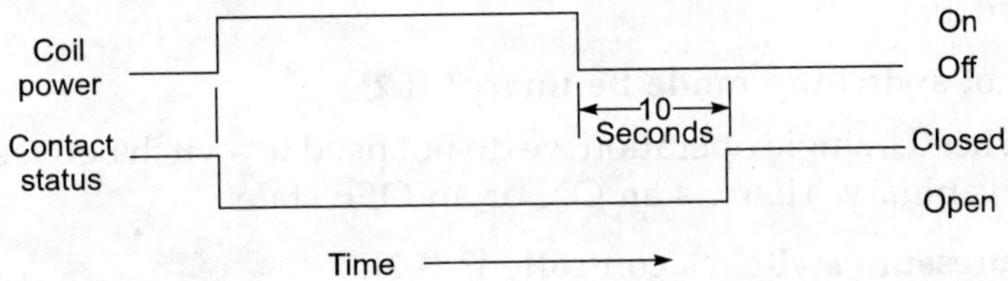


**105. Show diagrammatically the symbol of normally opened timed closed relay. (L2)**

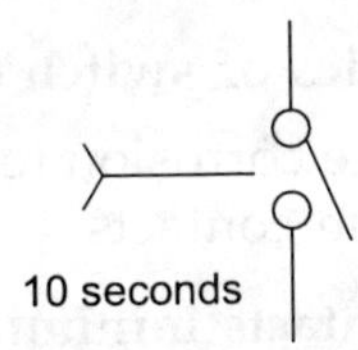


**Normally open-Timed closed relay**

**106. Name the different types of contactors (L2)**

- Air break contactors
- Vacuum contactors
- High voltage contactors

**107. Write a brief description on air break contactors. (L2)**

Air at atmospheric pressure surrounds the contacts of the contactor and extinguishes the arc when the current flowing through the circuit is interrupted. They are mostly used by low voltage motor control systems.

**108. On what basis should a relay be selected for an application? (L2)**

The following are to be looked into before selecting relays:

(i) Current rating – in the order of mA and A.
(ii) Voltage rating – 300 V to 600 V AC.
(iii) The switching types (high speed or low speed) should be decided properly based on the requirement.
(iv) Isolation between coil and contacts should be monitored properly.
(v) Type of contact, viz., NO or NC should be decided.
(vi) Types of protection like contact and coil protection should be noted.
(vii) Contact protection – to reduce arcing.
(viii) Coil protection – to reduce voltage surges.

**109. State any three important applications where relays are inevitable. (L2)**

(a) Relays can be used in low voltage circuits to reduce the noise level of the overall circuit during operation.
(b) They are also used in high voltage applications to reduce arcing.
(c) They can be used for providing time delays for opening and closing of contacts of switches.

**110. Why cannot switching mode be unary? (L2)**

When you need a single operation we do not need to switch between two. Switching operation is binary. There is an ON or an OFF state.

**111. How are pressure switches controlled? (L2)**

Pressure switches are controlled using mechanical force of piston, diaphragm, and liquid flow.

**112. What are the material characteristics of switch contacts? (L2)**

Electrically conductive, good surface corrosion resistance, inactive to oxidation are the material characteristics of switch contacts.

**113. How is arcing between switch contacts minimized? (L2)**

Arcing can be minimized by using snubber circuit in parallel to the switching contacts.

**114. What is wetting current? (L2)**

The minimum amount of current passed to keep the switch condition safe is the wetting current of the switch.

**115. Make-before-break contact is needed even though there is a break-before-make contact. Why? (L2)**

To enable the circuit even when the contact is in between two adjacent terminals we need make-before-break contact while break-before-make contact enables the circuit only at specific positions and not at intermediate positions (say between contacts 1 and 2).

**116. How can contact bouncing be minimized? (L2)**

1. The switching mechanism can be damped using oil or air.
2. Sliding design contacts can be used instead of direct contacts.
3. Controlling the movement of the pole by any means like any kinetic energy controlling device, contacts with slight variation in mass.
4. Using sealed contact mechanism wherein the movement of contacts can be controlled, thereby, their bounce can be controlled by any external means. For example, in the case of magnetic reed switch, the magnetic strip contacts are controlled by actuated permanent magnet externally, i.e., outside the sealed box and this kind of external control is highly feasible, cheap, reliable and accurate.

**117. What is the symbol of push button switch? (L2)**

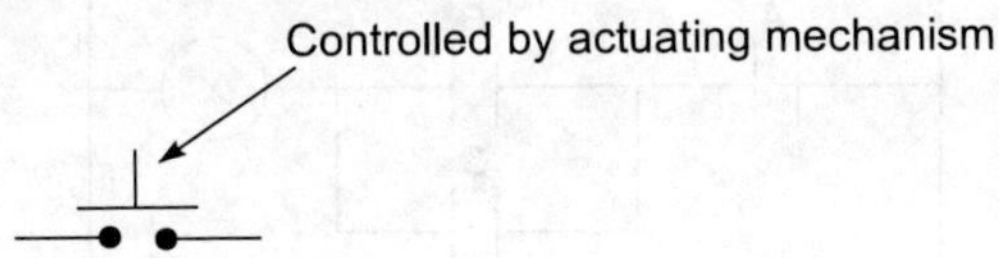


**118. What is the symbol of proximity switch? (L2)**

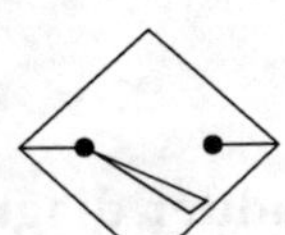

**119. Draw the symbol of speed switch. (L2)**

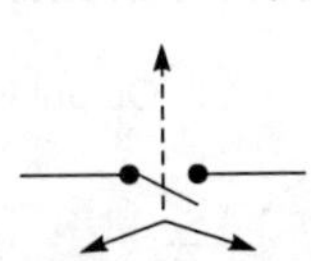

**120. Give examples for sealed contact switches. (L2)**

Examples of sealed contact switches are mercury switch, magnetic reed switch.

**121. What is a pole, and throw? (L2)**

Poles are the movable switch contacts and throws are the stationary switch contacts.

**122. Give a few examples for process switches. (L2)**

Pressure switch, temperature switch, level switch, speed switch are a few examples of process switches.

**123. What states of matter can be detected by level switches? (L2)**

Solid materials and liquid levels can be detected using level switches.

**124. Where can bouncing of switch contacts be seen clearly? (L2)**

Digital circuits with faster response time can clearly depict bouncing of switch contacts.

**125. There is something called Ladder Diagram frequently used with PLC. What is it and why is it named so? (L2)**

The control logics can be represented as ladder diagram which makes it easier for understanding the way the system works. Since it resembles a ladder it is named so. It is also referred as rung. A sample ladder diagram is shown below for reference.

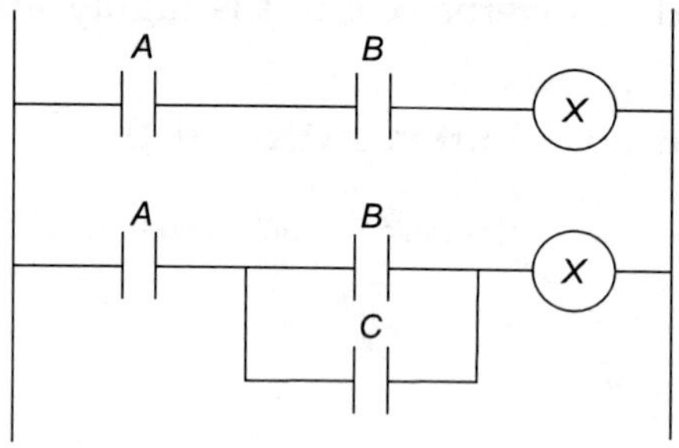


**126. Expand PLC. (L2)**

Programmable logic controller.

**127. What are the components of a ladder diagram? (L2)**

A ladder should have 2 power supplies (referred as power rails), horizontal lines which consist of input and output instructions and it is referred by the name rung. A sample diagram is presented below which has all these clearly indicated.

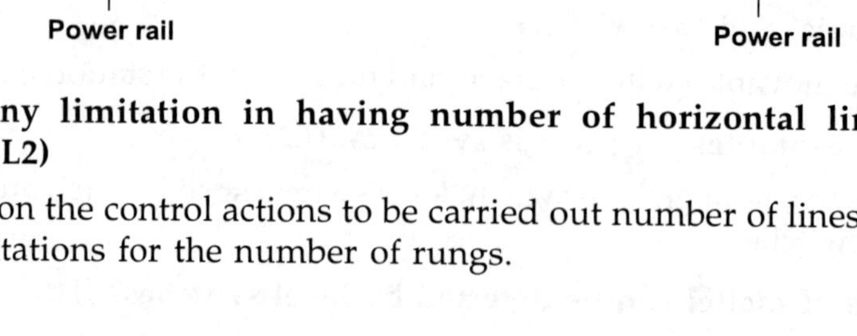


**128. Is there any limitation in having number of horizontal lines in a ladder diagram? (L2)**

No. Based on the control actions to be carried out number of lines may vary. There are no limitations for the number of rungs.

**129. What is a comparator and how does it function? (L2)**

As the name specifies, it is used to compare two signals. It compares two analog signals and gets a one-bit digital output.

**130. State one simple application where ADC is deployed. (L2)**

Simple reference could be mobile phones. Where, the user transmitted analog signal is getting digitized.

**131. Where is ADC and DAC used? Draw a schematic. (L2)**

The following simple schematic would explain the usage

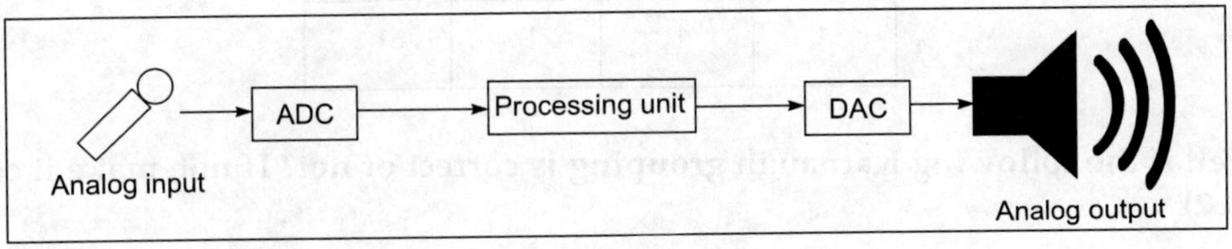


**132. Name an ADC that does not use a DAC in its design. (L2)**

The Slope ADC or the integrating ADC.

**133. The integrator in the slope ADC produces a ramp. What quantities does the ramp relate? (L2)**

Time and voltage are the two being related by the ramp.

**134. Slope ADC produces digital outputs at equal intervals of time. True or false? (L2)**

False. It provides outputs whenever the ramping voltage meets the analog signal voltage.

**135. How is the R/2NR DAC also known? (L2)**

It is also referred to as the binary-weighted-DAC.

**136. How will the resistances be in the R/2NR DAC? (L2)**

The resistances would be in the increasing order of powers of 2.

**137. The disadvantage of the R/2NR DAC is that it is not applicable for more than 3-bit binary numbers. True or false? (L2)**

False.

**138. The R/2NR DAC requires a wide variety of resistances for the design of the summing network. True or false? (L2)**

True.

**139. How many values of resistances does the R/2R DAC require? (L2)**

Two values. R and twice of R.

**140. Given below is the Karnaugh map table model taken for analysis. Why with the table get marked as 00011110 instead of 00 01 10 11 which being the correct order. (L3)**

Digital electronics is all about a bit being changed and here if the order is as 00 01 10 11 both the bits 01 get changed as 10 respectively. To avoid this, 00 01 11 10 is preferred as 01 gets changed only to 11 retaining 1 while changing 0.

| C \ AB | 00 | 01 | 11 | 10 |
|---|---|---|---|---|
| 0 | 0 | 1 | 1 | 1 |
| 1 | 1 | 1 | 1 | 1 |

**141. Tell if the following Karnaugh grouping is correct or not? If not, make it correct. (L2)**

*K* Map

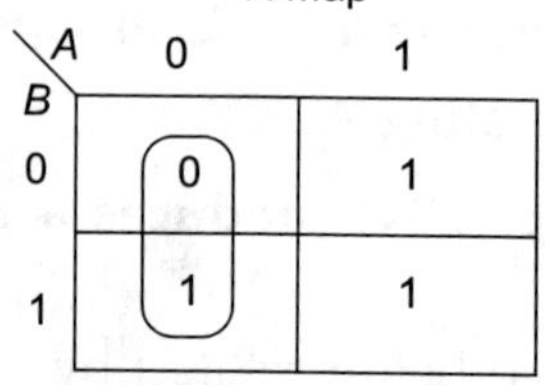


No, this is not proper as the grouping has not been done based on the Karnaugh map guidelines. So, the correct grouping is presented below.

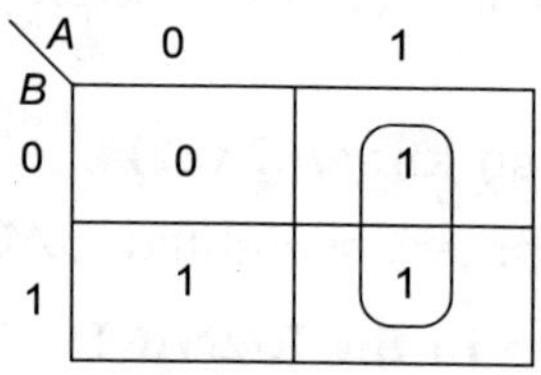


**142. Convey if the following grouping is correct or not? If not, provide the correct grouping. (L2)**

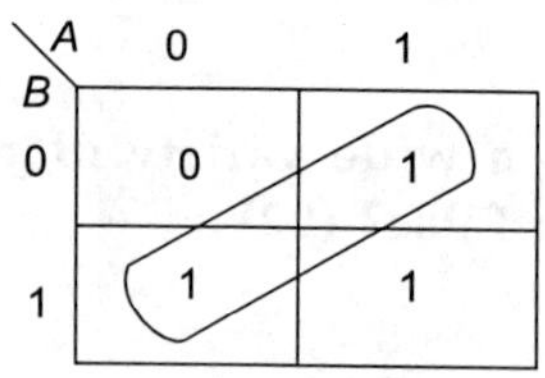

The above grouping is not correct. It cannot be diagonal. Hence, the correct grouping is one of the two methods as shown below.

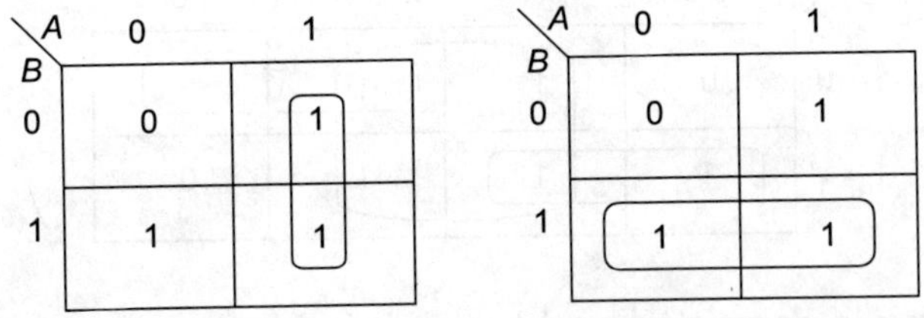


**143. Convey if the following grouping is correct or not? If not, provide the correct grouping. (L2)**

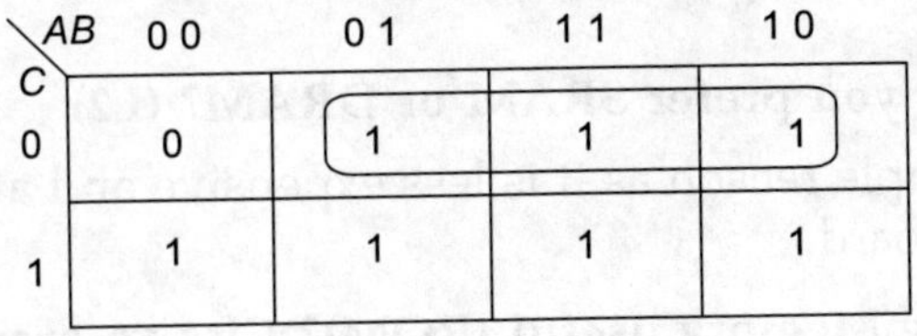


No. It is wrong grouping. Grouping should always contain elements in the order of powers of 2 (i.e., 20 = 1, 21 = 2, 22 = 4, 23 = 8, etc.,). Other orders of elements are not allowed and hence the given grouping is wrong.

**144. Convey if the following grouping is correct or not? If not, provide the correct grouping. (L2)**

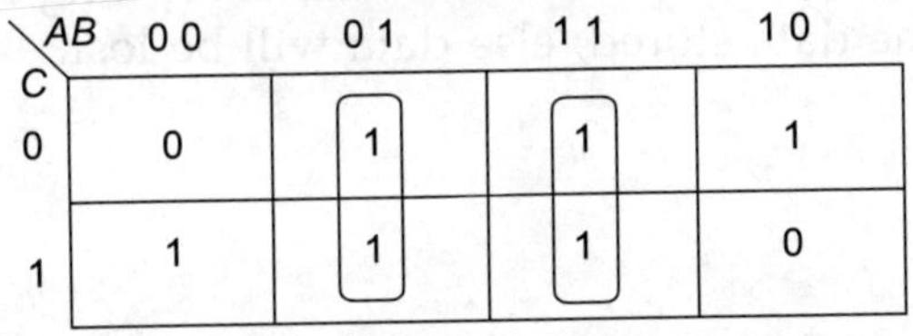


No. This is wrong. Grouping should be done as large as possible. If there is a chance for grouping 4 elements, it should not be preferred to group 2 elements. The correct grouping is presented below.

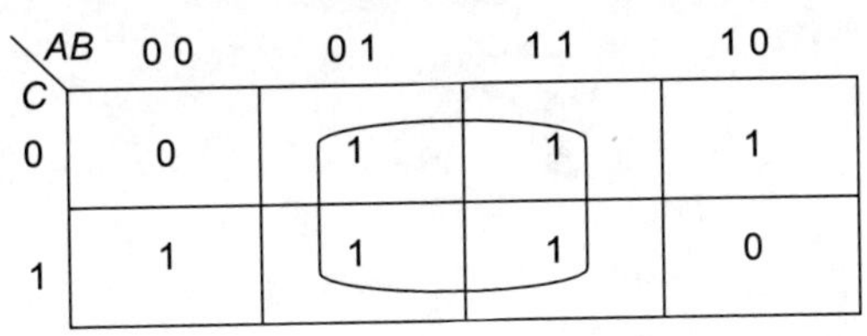

**145. Convey if the grouping below is correct. (L2)**

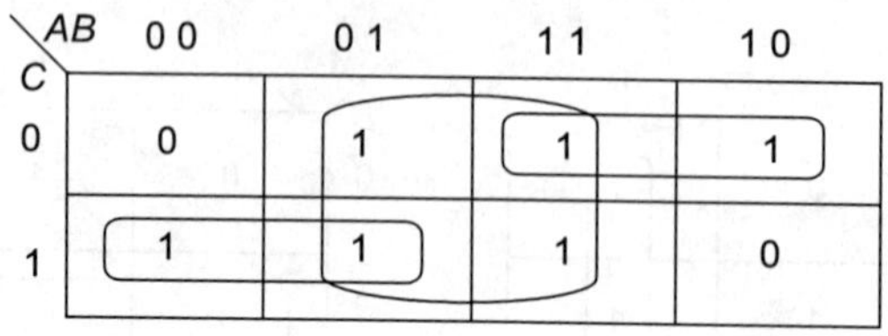


Yes. Overlapping is permitted and hence it is legal.

**146. What are the types of RAM available in the market? (L2)**

Static RAM, Dynamic RAM and SDRAM are the common types of RAM available for usage.

**147. Which one would you prefer SRAM or DRAM? (L2)**

DRAM, for the simple reason as it is less expensive and at the same time occupies less space in the board.

**148. How is Read Only Memory useful (ROM)? Cite an example. (L2)**

ROMs are very useful in building the embedded systems. Embedded systems should have some codes which are industry written and burnt and should not change over time. In this case ROMs are the only choice.

**149. What is volatility? Explain. (L2)**

Volatility is the tendency of the memory element to forget/lose all the stored information. For example, SRAMs are volatile. As long as the power is retained, SRAM will have the data stored, else data will be lost.

# 3 COMMUNICATION ENGINEERING

This communication engineering part covers the flow from basics to recent trends like optical communication, IoT, 5G and a few more. This helps the student to learn the communication part strongly which will be asked in many interviews. Different levels of difficulty are covered in all the areas so as to enhance the students' knowledge.

1. **What is Communication? (L1)**

   Communication denotes exchange of information, where a message is sent and received. It can be either half duplex, full duplex and simplex.

2. **What is channel? (L1)**

   To transmit and receive information, path is required. This path is called communication channel. It can be either a cable (fiber optic, twisted pair etc.) or broadcast (satellite, radio, etc.).

3. **What is half duplex? (L1)**

   We use a single path for communication. Since we use only path we can either transmit or receive at a time but not at the same time. Example: One-way traffic, Walkie-talkie.

4. **What is full duplex? (L1)**

   Full duplex means information can be sent and received in both the directions (from sender side and receiver side) at the same time. Example: Highways – on both sides vehicle can move at the same time.

5. **What is simplex communication system? (L1)**

   In simplex communication system, we either transmit or receive but we cannot do both at the same time. For example, in broadcast system only one transmitter is used to transmit information but one or many receivers will be used to receive it (news headlines scrolling in TV).

6. **What are the layers of communication? (L1)**

   - Application layer
   - Presentation layer

- Session layer
- Transport layer
- Network layer
- Data-link layer
- Physical layer

**7. Draw the block diagram of communication system? (L1)**

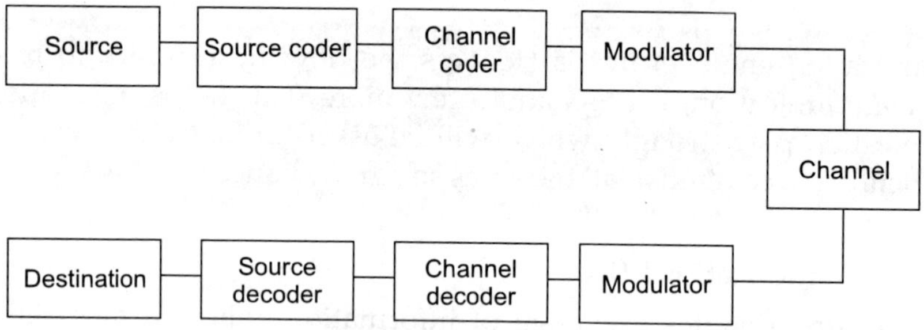


**8. What is Serial and parallel communication? (L1)**

Data can be transmitted between a sender and a receiver in two main ways:

- Serial
- Parallel

Serial communication is the method of transferring one bit at a time through a medium.

For example: Ethernet, RS-232, USB etc...

Parallel communication is the method of transferring blocks. For example, a BYTE of data is transmitted at the same time.

For example: Computer peripheral buses like: ISA, ATA, SCSI, PCI, RAM, etc.

**9. Define modulation. (L1)**

Modulation is a technique to transmit a signal over a long distance by varying properties of a waveform in a carrier signal.

**10. What is coding? What are the different types of coding? (L1)**

Messages are transmitted and received as bits. Based on the different applications we make the bits to fit into it accordingly. There are four different types of coding. They are:

- Data compression
- Error correction
- Cryptographic coding
- Line coding.

**11. What is data compression? (L1)**

Data compression means reducing the number of bits from the original size. It can be either lossless or lossy. Example: Zip file.

In lossless compression the data is received without any loss even after compressing it, which means file size is reduced but the content is same. Wherever we have had unwanted space, they would be removed.

In lossy compression, few bits are lost during compression. But, final information is still satisfactory.

**12. What is source coding and channel coding? (L1)**

Source coding means translation of information which are in the form of bits. It uses certain techniques to translate human understandable word into machine understandable bits like'0's or'1's. Some of the techniques are ASCII Codes, Huffman Codes, etc.

For example, the word "Computer" can be translated in the form of '0's and '1's by using either of these techniques.

Channel coding means error detection and correction techniques. When the message is sent and received, at the receiver end, the message is verified for errors during transmission (Reed Solomon codes)

**13. What are the two different types of coding techniques? (L1)**

- Liner block codes
- Convolution codes.

Liner block coding are of many types, They are:

- Cyclic codes
- Repetition codes
- Parity codes
- Polynomial codes
- Reed-Solomon codes, etc.

**14. Why antenna is required? (L1)**

Transmission and reception of signals in proper frequency can be done by the antennas. This transmission and reception of signals can be in one or multiple directions. This depends on the directivity of the antenna. The length of the antenna is decided by the frequency of transmission or reception.

**15. What is repeater? (L1)**

Whenever antennae are used to transmit signal and when the distance is large, repeaters are used so that they can listen to your signal and retransmit with large coverage area but with different frequency to avoid interference. Mostly, repeaters are mounted on top of the buildings, mountains to cover large area.

**16. What is base station? (L1)**

In simple words it is like host computer where many computers access details are maintained. Base station is where many antennas will be placed so that many users get connected to it.

**17. What is attenuation? (L1)**

When there is reduction in signal strength as it moves on, it is known as attenuation. This is because the signal is transferred through the medium and this medium will absorb some energy of the signal. In other words we say amplitude of the signal is reduced.

**18. What is path loss? (L1)**

When the distance is large and when the wave is travelling through the space for large distance then the strength of the signal becomes weak. This is due to the propagation distance. As we move on Line of Sight will not be there and this is the reason for the attenuation of the signal.

**19. What is meant by white Gaussian noise? (L1)**

A signal having all kinds of frequencies is termed a white signal. This is otherwise termed Gaussian noise. The reason for calling it a white signal is because noise has all range of frequencies.

**20. What is multiplexing? (L1)**

If we have many messages, and among that one message is selected and transmitted through single path or channel then it is called multiplexing.

**21. Different types of multiplexing? (L1)**

- Frequency division multiplexing
- Time division multiplexing
- Space division multiplexing
- Code division multiplexing
- Wavelength division multiplexing.

**22. What is a carrier? (L1)**

When we transmit the data we add weight to data so that it travels for long distance. This is done along with the input signal, i.e., when we want to travel short distance we go by walk but in case when we want to travel long distance we go by bus or train. Here bus or train is known as carrier signal.

**23. What is OFDM? (L2)**

It is defined as Orthogonal Frequency Division Multiplexing. Here the carriers are orthogonal to each other which help in avoiding interference.

**24. What is SNR? (L1)**

It is defined as Signal to Noise Ratio where the noise level should be less for better performance of the signal.

**25. What is multiple access technique? (L3)**

When the same spectrum or frequency is allowed to be used by many people or users simultaneously or at the same time then we use access scheme which is known as multiple access technique.

- CSMA/CD (Carrier Sense Multiple Access Collision Detection)
- CSMA/CA (Carrier Sense Multiple Access Collision Avoidance)

**26. What is frequency division multiplexing? (L1)**

In frequency division multiplexing (FDM) the given frequency or spectrum is divided into many channels where each user is allowed to use different channels. For example, when we use mobile phone and make a call while moving on road, the signal gets covered by shifting from one base station to other. At that time the base station allows us to use a channel, when we move on, only if the channel is free in other base station we can continue the call or else it is dropped. This is usually taken care by frequency which is allocated.

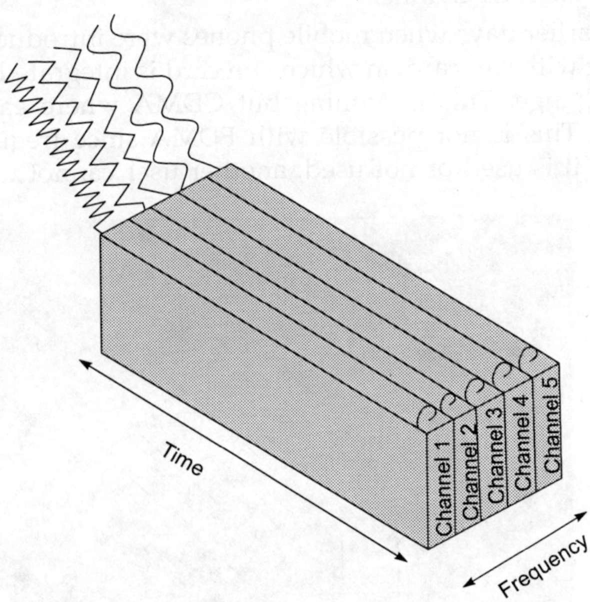


**27. What is time division multiplexing? (L1)**

When the allocated bandwidth or spectrum is not used by the user continuously then time division multiple access (TDMA) plays the role. Practically, user will not use the allocated channel all time, so when it is not used it remains idle, no other user cannot use it, i.e., when there is no continuous transmission, TDMA is used where all the channels can be used by all the users but for the allotted time or specific time period.

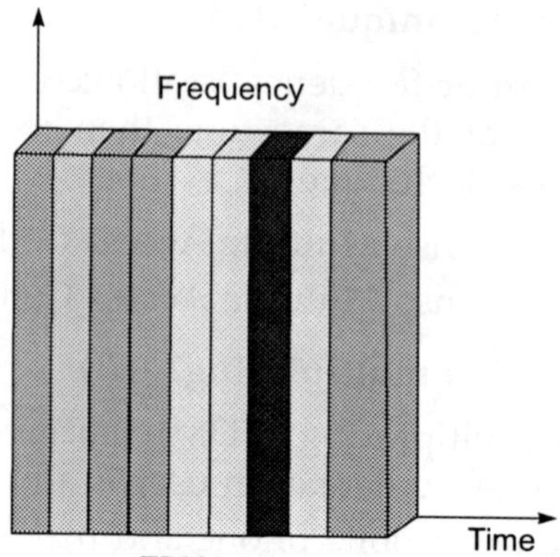

TDMA

**28. What is code division multiplexing? (L1)**

Here the frequency is shared between all channels at same time but individual code is given for each channel.

Example: In earlier days when mobile phones were introduced, Reliance marketed a phone along with sim card, in which sim card is integrated with the phone which you cannot change. This is nothing but CDMA where each user is allocated a specific code. This is not possible with FDMA since frequency is dedicated for users, though it is used or not used, another user cannot use it.

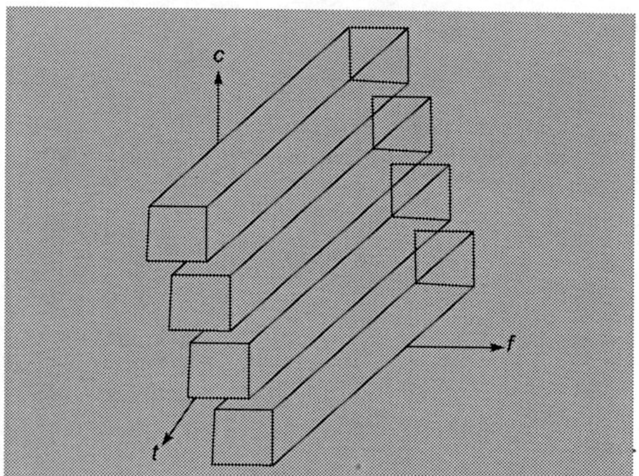

Code division multiplexing CDM

**29. What are the different generations of mobile phone? (L1)**

- 1G – Analog systems
- 2G – Digital systems (GSM)
- 3G – UMTS
- 3.5G – Mobile WiMAX
- 4G – LTE advanced (WiMAX).

**30. What is GSM? (L1)**

GSM – Global System for Mobile communication. GSM operates in the 900 MHz band (890 MHz – 960 MHz). It is developed by European Telecommunications Standards Institute (ETSI).

**31. Explain Bluetooth. (L1)**

Bluetooth is designed to be a personal area network, where participating entities are mobile and require sporadic communication with others. It is omnidirectional, i.e., it does not have line of sight limitation like infrared does. Bluetooth operates in the 2.4 GHz area of spectrum and provides a range of 10 metres. It offers transfer speeds of around 720 kbps.

**32. What is GPRS? (L1)**

General Packet Radio Service (GPRS) is a packet oriented mobile data service on the 2G and 3G cellular communication system's global system for mobile communications (GSM).

**33. Define EDGE? (L1)**

EDGE stands for enhanced data rates for GSM evolution which are used for connecting mobile with a packet data services. It operates in maximum speed range of 217.6 kbps Download & 108.8 kbps upload speed

**34. What is HSPA? (L2)**

It stands for High Speed Downlink Packet Access. It is modulated version of UMTS where downlink data speed is high when compared to UMTS (14 Mbps).

**35. What is LTE? (L2)**

It stands for Long Term Evolution. It has come up with the purpose of increasing capacity and speed compared to 3G. Maximum speed is 100 Mbps.

**36. What is uplink and downlink? (L1)**

When the communication happens from the ground to satellite it is called as uplink. Example: When we send photo or video to our friend it is known as uploading photos or documents. This is done by uplink.

When the communication happens from satellite to ground it is called as downlink Example: When the sent photo or video is received and viewed by our friend it is known as downloading the photo or video. This is done by downlink.

**37. What is handover? (L1)**

When we make a call through phone while driving vehicle or moving on road, it gets connected to one base station (BS). As we move on, the signal gives his hand to another base station so that the call is continued. This is known as handover. In simple handing over the signal from one BS to another BS is known as handover.

**38. What are the two types of hand over? (L1)**

- Hard handover
- Soft handover.

**39. What is hard and soft handover? (L2)**

In hard handover when the handover is happening it first breaks the connection from the corresponding base station and then makes connection to another cell base station. Simply, it is said as "break - before - make".

In soft handover when we move on from one cellular base station to another it first makes connection from the nearest base station to continue the call and then breaks the connectivity from the present base station. In simple it is termed "make- before-break".

**40. What is Shannon's Theorem? Why is it so important? (L2)**

This theorem gives an idea of how capacity of a system can be calculated when there is noise present in the signal. This calculation tells how much maximum bits can be received when we transmit *n* bits of information for *B* bandwidth.

$$C = B*\log2 (1 + S/N)$$

**41. Where do we use WDM? (L2)**

In optical fiber communication, when we try to inject photons inside fiber, as the core of the fiber is small it is difficult to inject, we divide the signal at different frequencies by wave division multiplexing so that it can be used for different frequencies though we inject signal at one frequency.

**42. What is Nyquist theorem? (L1)**

It states that when the bandwidth is B, it is able to receive 2B of data, ignoring noise. So practically, sometimes we ignore this theorem and we go for Shannon's capacity theorem where signal to ratio is considered.

**43. What is constellation diagram? (L1)**

It is used to represent the signal, e.g., if we consider BPSK scheme it has two bits '0' and '1' and in constellation diagram it is represented on X-axis. If we consider QPSK system, it has four bits so it is represented in circle format.

**44. What are the different types of networks? (L1)**

There are many kinds of networks; few of them are LAN, MAN, WLAN, CAN, WAN, etc.

**45. What is LAN? (L1)**

It stands for Local Area Network. It is used for communication within small areas like home, school, etc. The speed of the LAN ranges from 10 Mbps to 100 Mbps.

**46. What is mean by frame? (L2)**

It is used for flow of the bit. It will differentiate between the 2 different messages by using header and trailer. Frame has many segments.

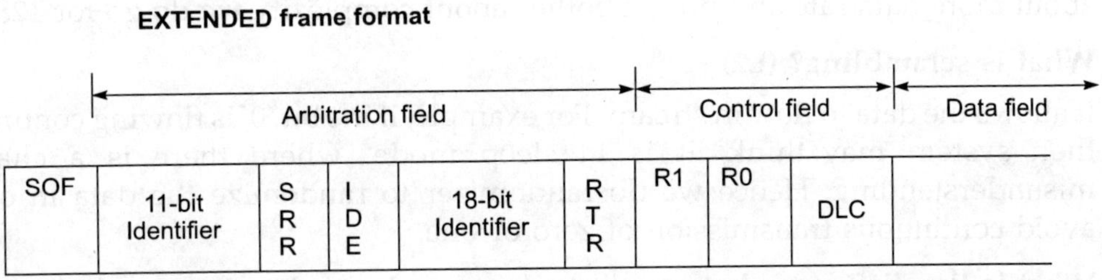


**47. What is CRC? (L2)**

It stands for Cyclic Redundancy Check. When we take 2 different sets of random bits, to the original message, it attaches a few bits which are calculated using some logic, for example, EX-OR between the random numbers in the transmitter side and in the receiver end we use same logic to remove these calculated bits. This is done to detect and correct the error. Sometimes it uses polynomial division technique to detect and correct errors.

**48. What is NRZ? (L2)**

It stands for Non-Return-to-Zero, where the transition happens by crossing zero level but will not stay in zero level to use the bandwidth efficiently. Here, logic '1' is set as positive value and logic '0' is set as negative value. This level crossing ensures continuous transmission.

**49. What is Shannon-Hartley theorem? (L2)**

It helps us to get the information about how much data can be transmitted per channel.

For example, how much water can be sent through a single pipe to fill the tank. Here pipe can be assumed as channel, water can be assumed as bits transmitted.

**50. What are the different modulation techniques? (L2)**

The most commonly used modulation techniques are PSK, FSK, ASK, QAM, OOK.

- PSK – Phase Shift Keying
- FSK – Frequency Shift Keying
- ASK – Amplitude Shift Keying
- QAM – Quadrature Amplitude Modulation (PSK + ASK)
- OOK – On-Off Keying.

**51. What modulation technique to be used at high frequency? (L3)**

As the frequency increases the data rate increases, so at high frequencies data rate is already high so simple modulation technique like OOK, BPSK is sufficient. High modulation technique makes the system complex.

**52. Why do we go for 64 or 128 QAM? (L3)**

Higher Modulation techniques are used to get higher data rate. But designing system at higher modulation technique involves more complexity. When we think about more data rate and do not bother about complexity we do go for 128 QAM.

**53. What is scrambling? (L2)**

It allows the data to flow in stream. For example, if the bit '0' is flowing continuously then system may think, it is in sleep mode, where there is a chance of misunderstanding. Hence we do randomizer to randomize the data in order to avoid continuous transmission of zero or one.

**54. What is the difference between base band and pass band transmission? (L2)**

When we transmit data without modulating, it means that, the signal strength is enough to transmit (includes zero frequencies). Such transmission is termed baseband transmission. Here, simulation takes less time. In passband transmission, we modulate the signal by shifting the frequency. So, zero frequency is not included in this. Such transmission is known as passband transmission. Here simulation takes little more time.

**55. What is MIMO? (L1)**

It means Multiple Input Multiple Output where multiple transmitters and receivers are there, which increases the capacity of the system.

**56. What is OFDM? (L1)**

It is defined as Orthogonal Frequency Division Multiplexing. Here, the carriers are orthogonal to each other which help in avoiding interference.

**57. What is PAPR? (L1)**

PAPR means Peak to Average Power Ratio. In OFDM we use many carriers and we add all these carriers at the receiver end. This makes the power to shoot up beyond the maximum power which leads to saturation of power amplifier.

**58. What is SNR? (L1)**

It is defined as Signal to Noise Ratio where the noise level should be less for better performance of the signal.

**59. What is the IEEE standard for wireless communication? (L2)**

It is 802.11/802.11a/802.11n etc. These standards vary, with respect to data rate and throughput.

**60. What is spread spectrum? (L2)**

The carrier signal is spread for the available bandwidth. Here the frequency of the spread signal is more than the original signal. The signal is spread by using a pseudorandom number which is scrambled so that at the receiver same pseudo-random code is used to remove it.

**61. What are the different types of spread spectrum? (L2)**

The different types of spread spectrum are:

- Direct Sequence Spread Spectrum
- Frequency Hopping Spread Spectrum.

**62. What is the role of data link layer? (L2)**

It is the layer in which electrical signals are converted into frame format (bit by bit format). This layer takes care of frame formation, where, synchronization, error detection and control, addressing, flow control, etc., are done.

**63. When congestion does occur? (L2)**

Congestion means traffic, when too many users attempt to use same frequency then congestion occurs.

**64. What protocol is used for assigning priority for transmission of bits? (L3)**

All communication systems use a protocol called CSMA/CD, which is given as Carrier Sense Multiple Access Collision Detection. Simpler terms when 2 messages try to transmit at the same time then collision occurs, wherein, by using this protocol it gets to know till what time it should wait and then re-transmit it again.

For example, consider three chairs and four persons. When everyone is moving around and trying to occupy the chairs, only three can successfully have their seat; the 4th person should try in the next round. Similarly, when music is 'on' (same musical chair scenario), all four persons can listen and when it is 'off', though everyone reacts, none can see how they reacted, but at the end when the chairs are occupied everyone will know which three have occupied and who will be trying next time. This is similar to collision detection.

**65. What are the responsibilities of presentation layer? (L2)**

Presentation layer deals with data compression, encryption, data translation, etc.

**66. What will happen when we use cable of more than required length? (L3)**

When we use cable of larger length it will result in signal loss because as they travel through the cable they tend to undergo losses like medium loss, type of material and losses made by it, as distance increases signal strength reduces.

**67. What is horizontal handover and vertical handover? (L4)**

In horizontal handoff, handshake happens between same technologies.

For example, handshake happens between same standards like WLAN.

In vertical handover, handshake happens between two different technologies.

For example, handshake happens between two different standards. That is WLAN and WMAN.

**68. What is line of sight and how does it work? (L2)**

Line of sight means communication that happens peer to peer, which in the presence of obstacle cannot transmit and receive data.

High frequencies can transmit and receive data only when they have line of sight. In simple words, it can be described as more secured because when you transmit you will see to whom you are transmitting and when you receive, you will see who has sent (same track).

**69. Define Booting? (L2)**

Booting is a primary process in which loads & starts the operating system. It is nothing but initialization of processor by executing of thousands of programming lines.

**70. What is boot loader? (L2)**

Boot loader is a set of programmes which is used to complete the booting process of a computer.

**71. What are Bios? (L2)**

BIOS stand for basic input/output system. It is a software program which is stored in computer's flash memory.

For example, when you turn on your computer the microprocessor will execute its first instruction; since the first set of instruction cannot be fetched from the hard disk directly, they are stored in BIOS which will help to complete the initial booting.

**72. Explain what happens when the computer is powered on? (L2)**

When a computer is powered on, the first process is booting the OS. However, the booting involves below given sequence.

- CPU is reset and registers are set to specific value.
- CPU jumps to address of BIOS (0xFFFF0).
- BIOS runs POST (Power-On Self-Test) and other necessary checks.
- BIOS jumps to MBR (Master Boot Record).
- Primary Boot loader runs from MBR and jumps to Secondary Boot loader.
- Secondary Boot loaders load Operating System.

**73. What is DHCP? (L2)**

DHCP stands for Dynamic Host Configuration Protocol. It is a network protocol which allows a particular server to assign an IP address automatically to a remote computer.

**74. What is DNS? (L2)**

DNS stands for Domain Name System which is a phone book for the Internet. It converts human readable host names to machine readable form.

For example: www.nokia.com is converted to address like 104.11.248.115 by the DNS which machine can understand.

**75. What is IP address? (L2)**

An Internet Protocol address (IP address) is a numerical name assigned to each computer. It is used to identify the network interface and also for locating the address.

IP address of the computer indicates the information we seek and where the information is available over the network.

For example, IP addresses look like 178.10.10.1 (IPv4), and 4551:ab8:0:5678:0:9101:5:1 (IPv6).

**76. What is TCP? (L2)**

TCP stands for Transmission Control Protocol. It's a machine language used by the computer to access the Internet. It manages the assembling of a message or file into smaller packets that are transmitted over the Internet and received by a TCP layer that reassembles the packets into the original message.

**77. What is MAC & MAC address? (L2)**

Media Access Control is a part of data link layer in OSI model which acts an interface between Data link layer & physical layer. It uses IEEE802 reference for addressing and channel access which allows several networks to communicate within a multiple access network.

MAC address is an alphanumeric character printed over the network card in the computer. It is used as unique identifier to identify the particular computer/device, e.g., 01-23-45-67-89-ab.

**78. What is the other name of MAC address? (L2)**

Burn In Address (BIA) is the other name of MAC address

**79. What is routing? (L2)**

It is a table which tells us how many packets (information) has been transferred and where it is exactly in the flow. It also helps us to add or delete packet.

**80. What are packets? (L2)**

When any file (e-mail message, HTML file, Graphics Interchange Format file, Uniform Resource Locator request, and so forth) is sent from one place to another on the Internet, the Transmission Control Protocol (TCP) layer of TCP/IP divides the file into "chunks" of an efficient size for routing. Each of these packets is separately numbered and includes the Internet address of the destination. The individual packets for a given file may travel different routes through the Internet. When they have all arrived, they are reassembled into the original file (by the TCP layer at the receiving end).

**81. Why do we do encryption? (L2)**

Encryption is like masking. For example, when we login to a Wi-Fi connection at homes, it will ask for our password which is encrypted. Only if you provide the password, you will be able to login. This password or code or retina or finger-print kind of protection is known as encryption so others cannot use without our knowledge.

**82. What are bit rate and baud rate? (L2)**

Bit rate is the number of bits per second that can be transmitted over a digital network.

Baud rate is the speed at which data is transmitted over a channel.

**83. What is IoT? (L2)**

IoT stands for Internet of Things. In simple words, we can say making objects/devices/things/speak to each other where both understand and complete a particular action. This communication happens with the Internet. This helps in reducing man work, provides safety, make places smart, etc.

**84. What is 5G? (L2)**

It is the fifth generation for communication. In 4G, it is Megabits per second transmission while in 5G it is Giga bit per second transmission, which is very challenging in accommodating users as well as increasing the speed/capacity of the system.

**85. What is VLC? (L3)**

It stands for Visible Light Communication. In all the discussions we usually study transmission and reception of signal, which happens through radio waves, but in VLC it happens through light waves, (i.e.) LI-FI (Light Fidelity).

**86. What is the standard followed by light wave communication and how does it work? (L3)**

It follows IEEE standard of 802.15.7. In transmission side we use light energy like LED and in reception side, we use photodiode. The modulating techniques, error detecting and correcting techniques is followed as like as radio waves communication system.

**87. Why do we go for four diode mixer? (L2)**

It has less loss and it can work in both cycles (positive and negative cycle).

**88. Why is inter-modulation bad? (L2)**

In inter-modulation we have high frequencies in the combination f1 + f2, f1 – f2, 2f, 3f, etc. This will affect the active devices like transistor which gets damaged because of different combination of frequencies, so inter-modulation is bad.

**89. What is millimeter wave? (L2)**

It is the high frequency wave which ranges from 30 GHz to 300 GHz. So it has Line of Sight where security is ensured, but because of its high frequency we can transmit data for shorter distance only.

**90. What is the network topology used in automobile systems? (L4)**

In automobile areas, they use CAN network which means Controller Area Network where different devices are made to communicate like sensors, actuators, etc.

**91. What is massive MIMO? (L2)**

When we use more than hundreds of antennas in a single system then it is termed Massive MIMO. When we use this kind of MIMO we will get a large number of paths with good diversity between each path but it is complex to place these many antennas very near without any interference.

**92. Give a practical application of Doppler's effect? (L2)**

It is used in detection of speed of a moving object, for example, the police vehicle measures the speed of the vehicle on road. There is stationary transmitter in the police vehicle with a defined frequency which shoots a wave at a moving object (for example, a car). The wave then hits the object and returns back. The receiver detects the returned wave and the amount of the Doppler shift in the return wave which will determine the speed of the vehicle.

**93. Define Clipping and its types. (L1)**

Clipping is a wave shaping technique in which an input waveform is clipped or cut in top half or bottom half or on both halves together to produce an output that resembles flattened version of the input. The electronic circuit which performs clipping operation is called a clipper.

Types of clipping:

- Positive
- Negative
- Biased
- Combinational

**94. Define Clamping. (L1)**

Clamping is a technique that places either the positive or negative peak of a signal above the desired input DC level. The circuit is called a clamper.

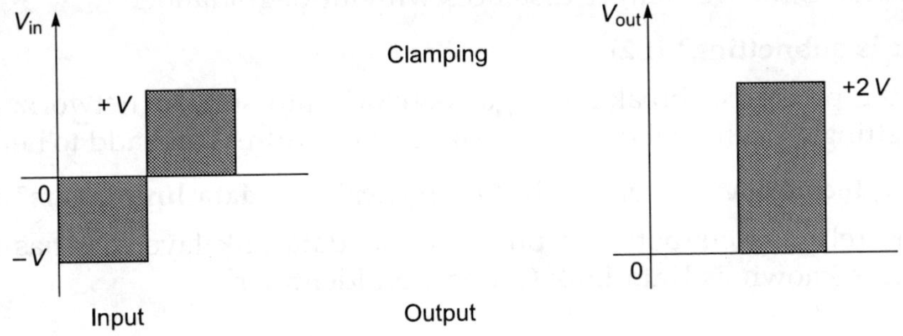

Types of clipping:

- ➢ Positive
- ➢ Negative

**95. What is VPN? (L4)**

VPN stands for Virtual Private Network. It allows an individual user to connect his business devices to a public network. Public networks are very secure networks. But VPN creates an encrypted connection over a less secure network.

For example, a person can use company laptop/gadget to connect to public Wi-Fi in railway stations, airports using VPN.

**96. What is Remote access? (L3)**

Remote access is a VPN technique by which a user can get an access to a computer or a network from a remote distance.

For example, when a system is kept power on in office connected to a network, the user can access the system from home via remote access option using another computer over the same network.

**97. What is Mobile Hotspot? (L2)**

Mobile hotspot uses Wi-Fi technique which provides connectivity between electronic gadgets within a particular distance. This is now widely used for wireless data transmission between mobile phones.

**98. What is sampling?**

The process of obtaining a set of samples from a continuous function of time $x(t)$ is referred to as sampling.

**99. State sampling theorem.**

It states that, while taking the samples of a continuous signal, it has to be taken care that the sampling rate is equal to or greater than twice the cut-off frequency and the minimum sampling rate is known as the Nyquist rate.

**100. What is a repeater?**

A repeater is an electronic device that receives a signal and retransmits it at a higher level and/or higher power, or onto the other side of an obstruction, so that the signal can cover longer distances without degradation.

**101. What is subnetting? (L2)**

It is the process of breaking larger network into smaller network. By using this subnetting we can steal the host position of ID address and add to network portion.

**102. Which technology covers both the physical and data link layers? (L3)**

Frame relay covers both the physical and data link layers. It has relay address which is known as Data Link Connection Identifier.

**103. What is meant by SSB? (L2)**

To transmit the modulated carrier wave only one side of band is used. Say if the total band is termed 'B' and left half of the band is termed 'b/2' and right half of the band is termed 'b/2' we use either of this 'b/2' to transmit the modulated carrier wave. This is known as 'Single Side Band' transmission.

**104. What is EDFA? (L3)**

EDFA stands for Erbium Doped Fiber Amplifier. It is used to amplify the optical signal.

**105. What is IoT protocol layer? (L4)**

This protocol layer has application layer, information processing layer, network construction layer, sensing and identification layer.

**106. What is ADSL? (L3)**

ADSL stands for Asymmetric Digital Subscriber Line which is used to transmit high frequency at the given power lines.

# 4 OPERATING SYSTEMS

This chapter deals with the basics of operating system and also includes some conceptual questions related to operating systems. At the end of the chapter you will be well versed in many operating system terms.

**1.** Operating system is a: (L1)
- (a) Application software
- (b) System software
- (c) Utility software
- (d) None of the above

**Answer: (b) System software**

**2.** Which of the following are operating systems: (L1)
- (a) Single and multi-tasking
- (b) Distributed and templated
- (c) Embedded and RTOS
- (d) All of the above

**Answer: (d) All of the above**

**3.** Operating system acts as an intermediary between: (L1)
- (a) Programs and hardware
- (b) User and application
- (c) Different applications
- (d) Multiple hardware

**Answer: (a) Programs and hardware**

**4.** Operating systems are found on which of the following devices: (L1)
- (a) Cellular phones
- (b) Video game console
- (c) Web-servers and supercomputers
- (d) All of the above

**Answer: (d) All of the above**

**5.** Which of the following is not a popular modern operating system: (L1)
- (a) Android
- (b) iOS
- (c) Linux and Windows
- (d) Microcontroller

**Answer: (d) Microcontroller**

**6.** A single-tasking system can run ________ program at a time, while a multitasking OS allows ________ program to be running in concurrency. (L1)
- (a) Many, one
- (b) One, one
- (c) one, many
- (d) many, many

**Answer: (c) one, many**

7. Multitasking may be characterized in _______ types: (L2)
   (a) Pre-emptive (b) co-operative
   (c) both (a) and (b) (d) neither (a) nor (b)

**Answer: (c) both (a) and (b)**

8. Unix like operating systems support: (L3)
   (a) Pre-emptive multitasking (b) co-operative multitasking
   (c) both (a) and (b) (d) neither (a) nor (b)

**Answer: (a) Pre-emptive multitasking**

9. Cooperative multitasking support is present in _______ versions of Microsoft windows: (L2)
   (a) 32-bit (b) 16-bit
   (c) 64-bit (d) 128-bit

**Answer: (b) 16-bit**

10. Single user operating systems have _______ facilities to distinguish users, but allow _______ programs to run: (L2)
   (a) No, multiple (b) All, no
   (c) Many, one (d) No, one

**Answer: (a) No, multiple**

11. _______ user operating system allows multiple users to interact with the system at the same time: (L2)
   (a) Single (b) Two
   (c) Multi-user (d) No

**Answer: (c) Multi-user**

12. A _______ operating system manages a group of distinct computers and makes them appear to be a single computer: (L2)
   (a) Single (b) Multi-user
   (c) Unix (d) Distributed

**Answer: (d) Distributed**

13. Distributed computations are carried out on: (L2)
   (a) Single machines (b) More than one machine
   (c) Two machines (d) Virtual machines

**Answer: (b) More than one machine**

14. When computers in a group work in _______, they form a distributed system: (L2)
   (a) Non-cooperation (b) independently
   (c) cooperation (d) virtually

**Answer: (c) cooperation**

**15.** Embedded operating systems are designed to be used in: (L2)
(a) Small machines like PDAs
(b) Supercomputers
(c) Laptops
(d) All of the above

**Answer: (a) Small machines like PDAs**

**16.** A RTOS is an OS that guarantees to process events or data in: (L3)
(a) Large amount of time
(b) One week
(c) Many years to complete
(d) Short amount of time

**Answer: (d) Short amount of time**

**17.** The earliest electronic digital systems had ________ operating systems: (L1)
(a) One
(b) No
(c) Many
(d) Few

**Answer: (b) No**

**18.** With the aid of the firmware and device drivers, the ________ provides the most basic level of control over all of the hardware devices: (L1)
(a) Application
(b) Kernel
(c) Memory
(d) User

**Answer: (b) Kernel**

**19.** ________ are central to operating systems: (L2)
(a) Interrupts
(b) Kernels
(c) Modes
(d) Applications

**Answer: (a) Interrupts**

**20.** The alternatives to interrupts are: (L1)
(a) Watchers
(b) Polling
(c) Other interrupts
(d) No alternatives

**Answer: (b) Polling**

**21.** Small stacks use ________ and large stacks use ________: (L2)
(a) Interrupts, Polling
(b) Polling, Polling
(c) None, Interrupts
(d) Polling, Interrupts

**Answer: (d) Polling, Interrupts**

**22.** When an interrupt is received, the computer's hardware: (L1)
(a) Continues doing current job
(b) Waits for user to suspend current program
(c) Automatically suspends current program and jumps to ISR
(d) Exits the program and goes to sleep state

**Answer: (c) Automatically suspends current program and jumps to ISR**

23. Interrupts may come from either the: (L1)
   (a) Computer hardware (b) Running program
   (c) Neither (a) nor (b) (d) Either (a) or (b)

**Answer: (d) either (a) or (b)**

24. Interrupts are handled by the operating system's: (L1)
   (a) User (b) Kernel
   (c) Application (d) Hardware

**Answer: (b) Kernel**

25. Modern CPUs support _______ modes of operation: (L1)
   (a) Single (b) Two
   (c) Multiple (d) No

**Answer: (c) Multiple**

26. Which of the following are valid modes of operation: (L1)
   (a) Protected mode (b) Supervisor mode
   (c) Application mode (d) Both (a) and (b)

**Answer: (d) Both (a) and (b)**

27. The _______ mode is used by the operating system's kernel for low level tasks that need unrestricted access to hardware: (L2)
   (a) Supervisor (b) Application
   (c) User (d) Kernel

**Answer: (a) Supervisor**

28. Applications operate within _______ mode, and can only use hardware by communicating with the kernel, which controls everything in supervisor mode: (L3)
   (a) Supervisor (b) Protected
   (c) User (d) All of the above

**Answer: (b) Protected**

29. When a computer first starts up, it is automatically running in _______ mode: (L1)
   (a) No (b) User
   (c) Supervisor (d) Protected

**Answer: (c) Supervisor**

30. A multiprogramming operating system kernel must be responsible for managing all _______ which is currently in use by programs: (L1)
   (a) Users (b) Time
   (c) System Memory (d) Interrupts

**Answer: (c) System Memory**

31. By ________ memory addressing, the OS can use the same memory locations for multiple tasks: (L1)
   (a) Virtual (b) Real
   (c) Imaginary (d) Offset

**Answer: (a) Virtual**

32. Virtual memory provides the user with the perception that there is a much larger amount of ________ in the computer than is really there: (L1)
   (a) ROM (b) Data
   (c) Tertiary memory (d) RAM

**Answer: (d) RAM**

33. ________ refers to the running of multiple independent computer programs on the same computer: (L1)
   (a) Single tasking (b) Multitasking
   (c) Both (a) and (b) (d) Neither (a) nor (b)

**Answer: (b) Multitasking**

34. Kernel contains a ________ program which determines how much time each process spends executing and in which order control should be passed to programs: (L1)
   (a) Application (b) Scheduling
   (c) User (d) All of the above

**Answer: (b) Scheduling**

35. An early model which governed the allocation of time to programs was called ________: (L1)
   (a) Cooperative multitasking (b) Scheduling
   (c) Multithreading (d) None of the above

**Answer: (a) Cooperative multitasking**

36. The so called passing of control between the kernel and applications is called a ________: (L1)
   (a) Thread switch (b) Task switch
   (c) Process switch (d) Context switch

**Answer: (d) Context switch**

37. Context switches are usually ________, and much of the design of operating systems is to optimize the use of context switches: (L1)
   (a) Long (b) Computationally easy
   (c) Computationally intensive (d) None of the above

**Answer: (c) Computationally intensive**

**38.** Potential trigger for a context switch is: (L2)

(a) Multitasking (b) Interrupt handling

(c) Mode switching (d) All of the above

**Answer: (d) All of the above**

**39.** Which multitasking can hang the entire system if it enters an infinite loop: (L2)

(a) Cooperative (b) Preemptive

(c) Independent process scheduler (d) All of the above

**Answer: (a) Cooperative**

**40.** Which multitasking ensures that all programs are given regular time on the CPU: (L1)

(a) Non-preemptive (b) Preemptive

(c) Independent process scheduler (d) All of the above

**Answer: (b) Preemptive**

**41.** Computers store data on disks using: (L1)

(a) Database (b) Computer storage

(c) Soft copy (d) Files

**Answer: (d) Files**

**42.** The specific way in which files are stored on a disk is called a: (L1)

(a) Database (b) Storage system

(c) File system (d) All of the above

**Answer: (c) File system**

**43.** Files allows data to be stored in a hierarchy of directories or folders arranged in a: (L2)

(a) Graph (b) Database

(c) Binary tree (d) Directory tree

**Answer: (d) Directory tree**

**44.** A ________ switch is an abstraction layer on top of a more concrete file system: (L3)

(a) Virtual file system (VFS) (b) Regular file system

(c) Real file system (d) All of the above

**Answer: (a) Virtual file system (VFS)**

**45.** The purpose of a VFS is to allow client applications to access different types of concrete file systems in a ________ way: (L2)

(a) Distributed (b) Uniform

(c) Both uniform and distributed (d) Random

**Answer: (b) Uniform**

46. A connected storage device, such as a hard drive, is accessed through a: (L1)
    (a) User driver (b) Device driver
    (c) Application driver (d) All of the above

**Answer: (b) Device driver**

47. A device driver is a specific type of computer which allows interaction with: (L1)
    (a) Hybrid devices (b) Software devices
    (c) Hardware devices (d) None of the above

**Answer: (c) Hardware devices**

48. The key design goal of device drivers is: (L1)
    (a) Virtualise (b) User preference
    (c) Hardware devices (d) Abstraction

**Answer: (d) Abstraction**

49. The function of the ________ is to translate operating system mandated function calls into device specific calls: (L1)
    (a) Application driver (b) Device driver
    (c) User driver (d) All of the above

**Answer: (b) Device driver**

50. Computers running dissimilar operating systems can participate in a common ________ for sharing resources: (L1)
    (a) Network (b) Storage
    (c) Web (d) All of the above

**Answer: (a) Network**

51. Networking allows a program on a computer, called a ________, to connect via a network to another computer, called a ________: (L1)
    (a) Client, Server (b) User, Application
    (c) Source, Destination (d) Producer, Consumer

**Answer: (a) Client, Server**

52. A modern operating system provides access to a number of resources and needs a: (L1)
    (a) Hardware (b) User
    (c) Security (d) Application

**Answer: (c) Security**

53. Every computer that is to be operated by an individual requires a: (L1)
    (a) System interface (b) User interface
    (c) No interface (d) Library interface

**Answer: (b) User interface**

54. The user interface is usually referred to as a ________ and is essential if human interaction is to be supported: (L1)
   (a) Shell
   (b) Application
   (c) Kernel
   (d) Software

**Answer: (a) Shell**

55. Which of the following are input hardware devices: (L1)
   (a) Keyboard
   (b) Mouse
   (c) Card reader
   (d) All of the above

**Answer: (d) All of the above**

56. The most common forms of a user interface have historically been the: (L1)
   (a) Command line interface
   (b) Graphical user interface
   (c) Both (a) and (b)
   (d) Neither (a) nor (b)

**Answer: (c) Both (a) and (b)**

57. A real-time operating system is an operating system intended for applications with: (L1)
   (a) Fixed deadlines
   (b) Random deadlines
   (c) No deadline
   (d) Multiple deadlines

**Answer: (a) Fixed deadlines**

58. The CPU can load instructions only from ________, so any programs to run must be stored there: (L1)
   (a) Keyboard
   (b) Memory
   (c) Mouse
   (d) Monitor

**Answer: (b) Memory**

59. The main requirement for secondary storage is to be able to hold large quantities of data: (L1)
   (a) Never erasable thereafter
   (b) Few seconds
   (c) Temporarily
   (d) Permanently

**Answer: (d) Permanently**

60. The goals of the operating system are: (L1)
   (a) Execute user programs and make solving user problems easier
   (b) Make the computer system convenient to use
   (c) Use the computer hardware in an efficient manner
   (d) All of the above

**Answer: (d) All of the above**

**61.** Operating system is a: (L1)

(a) Resource allocator
(b) Control program
(c) Both (a) and (b)
(d) Neither (a) nor (b)

**Answer: (c) Both (a) and (b)**

**62.** I/O devices and the CPU can execute: (L1)

(a) At different times only
(b) Concurrently
(c) On sleep state
(d) Based on time and space

**Answer: (b) Concurrently**

**63.** Interrupt transfers control to the interrupt service routine generally, through the: (L1)

(a) Interrupt vector
(b) Scalar
(c) Stack
(d) Interrupt service routine

**Answer: (a) Interrupt vector**

**64.** Interrupt architecture must save the ________ of the interrupted instruction: (L1)

(a) Address
(b) Data
(c) Control
(d) Time

**Answer: (a) Address**

**65.** A ________ is a software generated interrupt: (L1)

(a) Mask
(b) Router
(c) Trap
(d) Timer

**Answer: (c) Trap**

**66.** An operating system is ________: (L1)

(a) Data driven
(b) Interrupt driven
(c) Architecture driven
(d) Time driven

**Answer: (b) Interrupt driven**

**67.** Main memory is ________ storage device: (L1)

(a) Data
(b) Not
(c) Non-volatile
(d) Volatile

**Answer: (d) Volatile**

**68.** A ________ is a computer application for controlling unattended background program execution: (L1)

(a) Software
(b) Job scheduler
(c) Task
(d) Compiler

**Answer: (b) Job scheduler**

69. Processes are ________ in and out of main memory to the disk: (L1)
(a) Swapped
(b) Maintained
(c) Closed
(d) Discontinued

**Answer: (a) Swapped**

70. Operating system and the users share ________ resources of the computer system: (L1)
(a) Firmware
(b) Software
(c) Hardware
(d) Both (b) and (c)

**Answer: (d) Both b and c**

71. A bit, called the ________ is added to the hardware of the computer to indicate the current mode: kernel (0) or user (1): (L2)
(a) User bit
(b) Kernel bit
(c) Mode bit
(d) Application

**Answer: (c) Mode bit**

72. With the mode bit, we are able to distinguish between a task that is executed on behalf of the ________ and one that is executed on behalf of the ________: (L1)
(a) Hardware, Software
(b) Operating system, User
(c) Kernel, Hardware
(d) Application, Thread

**Answer: (b) Operating system, User**

73. A user application requests a service from the operating system via a ________: (L1)
(a) System call
(b) User call
(c) Kernel call
(d) Thread call

**Answer: (a) System call**

74. At system boot time, the hardware starts in ________: (L1)
(a) Application mode
(b) User mode
(c) Kernel mode
(d) No mode

**Answer: (c) Kernel mode**

75. Operating system is loaded and starts user applications in ________ mode: (L1)
(a) Kernel mode
(b) User mode
(c) Application mode
(d) No mode

**Answer: (b) User mode**

76. The hardware allows privileged instructions to be executed only in ________: (L1)
(a) Kernel mode
(b) User mode
(c) Application mode
(d) No mode

**Answer: (a) Kernel mode**

77. The dual mode of operation provides us with the means for ________ the operating system from errant user: (L1)
   (a) Releasing (b) Protecting
   (c) Exposing (d) Nothing to

**Answer: (b) Protecting**

78. Example of a privileged instruction is ________: (L1)
   (a) I/O control (b) Timer management
   (c) Interrupt management (d) All of the above

**Answer: (d) All of the above**

79. When a system call is executed it is treated by the hardware as a ________: (L1)
   (a) Low priority call (b) Hardware interrupt
   (c) Software interrupt (d) Thread

**Answer: (c) Software interrupt**

80. Additional information needed for the interrupt request may be passed in ________: (L1)
   (a) Registers on stack (b) Memory
   (c) Both (a) and (b) (d) Neither (a) nor (b)

**Answer: (c) Both (a) and (b)**

81. ________ has no mode bit and therefore no dual mode: (L2)
   (a) MS-DOS written for Intel 8088 architecture
   (b) PC DOS
   (c) ROM-DOS
   (d) Free DOS

**Answer: (a) MS-DOS written for Intel 8088 architecture**

82. To accomplish the goal of user program not to get stuck in infinite loop or to fail to call system services and never return control to operating system, we use ________: (L1)
   (a) Crystal (b) Timer
   (c) Memory (d) Lock

**Answer: (b) Timer**

83. A timer can be set to interrupt the computer after a ________ period: (L1)
   (a) Thread end (b) Day
   (c) Specified (d) Variable

**Answer: (c) Specified**

84. A variable timer is generally implemented by a _______ and a _______: (L1)
(a) Variable clock, Timer
(b) Timer, Counter
(c) Watch, Clock
(d) Fixed rate clock, Counter

**Answer: (d) Fixed rate clock, Counter**

85. The _______ sets the counter to decrement on every clock tick. (L1)
(a) Operating system
(b) User application
(c) Utility software
(d) Hardware

**Answer: (a) Operating system**

86. A process needs _______ resources to accomplish its tasks (L1)
(a) CPU time
(b) Memory
(c) Files and I/O devices
(d) All of the above

**Answer: (d) All of the above**

87. Resources are given to the process when it is _______ (L1)
(a) Running
(b) Created
(c) Both (a) and (b)
(d) Neither (a) nor (b)

**Answer: (c) Both (a) and (b)**

88. When the process terminates, the operating system will _______ any reusable resources. (L1)
(a) Create
(b) Reclaim
(c) Not reclaim
(d) Destroy

**Answer: (b) Reclaim**

89. A program is a _______ entity while a process is a _______ entity. (L1)
(a) Active, Passive
(b) Passive, Active
(c) Useless, Useful
(d) Negative, Positive

**Answer: (b) Passive, Active**

90. A single threaded process has _______ program counter specifying the next instruction to execute. (L1)
(a) One
(b) Two
(c) Three
(d) Many

**Answer: (a) One**

91. The execution of a single threaded process must be _______ (L1)
(a) Alternative
(b) Random
(c) Sequential
(d) None of the above

**Answer: (c) Sequential**

**92.** A multithreaded process has ________ program counters, each pointing to the next instruction to execute for a given thread. (L2)

(a) Single (b) Multiple
(c) Two (d) No

**Answer: (b) Multiple**

**93.** Which of the following are system processes (L1)

(a) Operating system processes (b) User processes
(c) Both (a) and (b) (d) Neither (a) nor (b)

**Answer: (c) Both (a) and (b)**

**94.** Operating system processes and system processes can potentially execute (L1)

(a) In sequence only (b) In random fashion
(c) Concurrently (d) In sleep state

**Answer: (c) Concurrently**

**95.** In connection with process management, OS is responsible for: (L1)

(a) Scheduling processes and threads on the CPUs
(b) Creating and deleting both user and system processes
(c) Suspending and resuming processes
(d) All of the above

**Answer: (d) All of the above**

**96.** Main memory is a repository of quickly accessible data shared by the (L1)

(a) CPU and I/O devices
(b) Primary memory and secondary memory
(c) Cache and memory
(d) None of the above

**Answer: (a) CPU and I/O devices**

**97.** The computer has single storage system for storing data as well as program to be executed as per ________ architecture. (L1)

(a) Harvard (b) Von-Neumann
(c) Any computer (d) Split cache architecture

**Answer: (b) Von-Neumann**

**98.** ________ is not possible with Von-Neumann architecture (L1)

(a) Programming (b) Application
(c) Sequencing (d) Pipelining

**Answer: (d) Pipelining**

99. As per Harvard architecture, computer has two separate memories for storing ____ (L1)
   (a) Multiple data (b) Many programs
   (c) Data and program (d) Address and data

**Answer: (c) Data and Program**

100. Number of stages in the pipeline is ________ (L1)
   (a) Fixed (b) Varies from system to system
   (c) Runtime configurable (d) None of the above

**Answer: (b) Varies from system to system**

101. In connection with file management, OS is responsible for ________ (L1)
   (a) Creating and deleting files
   (b) Creating and deleting directories to organize files
   (c) Mapping files onto secondary storage
   (d) All of the above

**Answer: (d) All of the above**

102. In connection with disk management, OS is responsible for ________ (L1)
   (a) Free space management (b) Storage allocation
   (c) Disk scheduling (d) All of the above

**Answer: (d) All of the above**

103. Information normally kept in some storage system, when used is copied into a faster storage system called ________ (L1)
   (a) Compact disc (b) Caching
   (c) Secure digital card (d) RAM

**Answer: (b) Caching**

104. Consistency of shared resource data that ends up stored in multiple local caches is called ________ (L2)
   (a) Cache coherency (b) Non-caching
   (c) Memory barrier (d) RAM

**Answer: (a) Cache coherency**

105. Only the ________ knows the peculiarities of the specific device to which it is assigned. (L1)
   (a) Hardware (b) Cache
   (c) Device driver (d) Application

**Answer: (c) Device driver**

**106.** _______ is any mechanism for controlling the access of processes or users-to the resources defined by a computer system. (L1)

(a) Software (b) Hardware
(c) Device driver (d) Protection

**Answer: (d) Protection**

**107.** The job of "security" to defend a system from (L1)

(a) Viruses and worms (b) Denial-of-service attacks
(c) Identity theft (d) All of the above

**Answer: (d) All of the above**

**108.** A _______ is a collection of physically separate, possibly heterogeneous, computer systems (L1)

(a) Software system (b) Server
(c) Distributed system (d) All of the above

**Answer: (c) Distributed system**

**109.** A _______ in the simplest terms is a communication path between two or more systems. (L1)

(a) Application (b) Software
(c) Network (d) Kernel

**Answer: (c) Network**

**110.** Distributed systems depend on _______ for their functionality. (L1)

(a) Networking (b) Software
(c) Hardware (d) Application

**Answer: (a) Networking**

**111.** In computing, a _______ is a computer program that operates or controls a particular type of device that is attached to a computer. (L1)

(a) Kernel (b) Software
(c) Application (d) Device Driver

**Answer: (d) Device driver**

**112.** A driver provides a _______ to hardware devices, enabling operating systems and other computer programs to access hardware functions (L1)

(a) Application interface (b) Software interface
(c) Hardware interface (d) No interface

**Answer: (b) Software interface**

**113.** A driver typically communicates with the device through the ________ to which the hardware connects. (L1)

(a) Network (b) Application
(c) Computer bus (d) Software

**Answer: (c) Computer bus**

**114.** A ________ provides unbuffered, direct access to the hardware device. (L1)

(a) Character device (b) Block device
(c) Both (a) and (b) (d) Neither (a) nor (b)

**Answer: (a) Character device**

**115.** Block devices provide ________ access to the hardware, such that the hardware characteristics of the device are not visible. (L1)

(a) Network (b) Buffered
(c) Random (d) Aligned data

# 5 C PROGRAMMING

These questions from C Programming shall help the reader to understand the basics of C Programming and to crack the questions in this area. Initially, it might be tough, but second time it would be easier.

1. What is the output of the code? (L4)

```
main()
{ int a,b;
  printf("enter two int");
  scanf("%d %d",&a,&b);
  printf("%d",a, "%d",b); }
```

(a) The value given for 'a' and 'b' will be printed.
(b) The value of 'a' alone will be printed.
(c) The value of 'b' alone will be printed.
(d) The value given for 'a' along with 0.

**Answer: (b) Once the value given for 'a' is printed in the console the printf statement is terminated. It doesn't return back to printf again for getting 'b'. So gives only 'a'.**

2. What happens on running the code? (L4)

```
main()
{ int a,b;
  printf("enter two int");
  scanf("%d" "%d",&a,&b);
  printf("%d %d",a,b);   }
```

(a) The value given for 'a' and 'b' will be scanned and core dumps.
(b) The value of 'a' alone will be scanned and core dumps.
(c) The value of 'b' alone will be scanned and core dumps.
(d) The value given for 'a' and 'b' will be got and the same will be printed.

**Answer: (d) The two format specifiers even though given in two separate quotes it is not separated by a comma. Hence, they are treated as a single argument and works normally.**

3. What happens on running the code? (L4)

```
main()
{ int a,b;
  printf("enter two int");
  scanf("%d%d",&a,&b);
  printf("%d " "%d",a,b); }
```

(a) The value given for 'a' and 'b' will be scanned and core dumps.
(b) The value given for 'a' and 'b' will be scanned. The value of 'a' alone will be scanned and core dumps.
(c) The value given for 'a' and 'b' will be scanned. The value of 'b' alone will be scanned and core dumps.
(d) The value given for 'a' and 'b' will be got and the same will be printed.

**Answer: (d) The two format specifiers even though given in two separate quotes it is not separated by a comma. Hence, they are treated as a single argument and works normally.**

4. What is the output of the code? (L4)

```
main()
{ int a=9;
  printf("%D ",a); }
```

(a) The value given for 'a' and 'b' will be printed.
(b) error
(c) the value given for 'a' along with 0
(d) %D

**Answer: (d) format specifiers are case sensitive.**

5. What is the output of the code? (L4)

```
main()
{ int a,b;
  printf("enter an int");
  scanf("%d",&a,&b);
  printf("%d %d",a,b); }
```

(a) The value given for 'a' will be printed twice.
(b) error
(c) core dumps
(d) the value given for 'a' along with 0

**Answer: (d) The given value is taken to 'a' while b is initialised by default to 0.**

6. What is the output of the code? (L3)

```
main()
{ int a,b;
  printf("enter two int");
  scanf("%d %d",&a,&b);
  printf("%d",a,b); }
```

(a) The value given for 'a' will be printed twice.
(b) The value of 'a' alone
(c) The value of 'b' alone
(d) The value given for 'a' along with 0

**Answer: (b) One format specifier with two arguments takes the first argument value to output stream neglecting the second.**

7. What is the output of the code? (L3)

```
main()
{ int a,b;
  printf("enter an int");
  scanf("%d",&a,&b);
  printf("%d %d",a,b);}
```

(a) The value given for 'a' will be printed twice.
(b) Error
(c) Core dumps
(d) The value given for 'a' along with 0

**Answer: (d) The given value is taken to 'a' while 'b' is initialised by default to 0.**

8. What is the output of the code? (L4)

```
main()
{ int a,b;
  printf("enter two int");
  scanf("%d",&a,"%d",&b);
  printf("%d %d",a,b); }
```

(a) The value given for 'a' and 'b' will be printed.
(b) Error
(c) Core dumps
(d) The value given for 'a' along with 0

**Answer: (d) Once the value given for 'a' is taken to memory the scanf statement is terminated. It doesn't return back to the scanf again for getting 'b'. So gives 0 for 'b'.**

9. What is the output of the code? (L2)

```
void main()
{ int g;
  printf("%x",g); }
```

(a) Garbage_value (b) 0
(c) 0x00 (d) Error

**Answer: (c) Even though the format specifer is %x the intial value 0 is not printed in hexadecimal format.**

10. What is the output of the code? (L1)

```
main()
{ printf("%o",76); }
```

(a) 76 (b) Garbage value
(c) 076 (d) 114

**Answer: (d) The format specifier takes the corresponding value without any issues. As it is in int it is converted into its corresponding octal number.**

11. What is the output of the code? (L2)

```
main()
{ printf("%o",76.345); }
```

(a) 0x76 (b) 114
(c) Garbage value (d) 76

**Answer: (c) The format specifier takes the corresponding value without any issues. As it is in float it tries to converted into its corresponding hexadecimal, but results in garbage values as float does not have octal value.**

12. What will be the size of the following program in a 64-bit processor with gcc compiler? (L1)

```
void main()
{ int a,b;
  double c;
  printf("enter the two numbers");
  scanf("%d %d",&a,&b);
  c=a/b;
  printf("their quotient is %lf",c); }
```

(a) 4 bytes (b) 8 bytes
(c) 16 bytes (d) Input dependent

**Answer: (c) Irrespective of the input an int variable takes 4 bytes and double takes 8 bytes. So the total is 4+4+8=16 bytes. It is also to be noted that the other functions doesn't consume any space because function only act upon the values that are stored.**

13. What will be the size of the following program in a 64-bit processor with gcc compiler? (L2)

```
void main()
{ a,b;
  printf("enter the two numbers");
  scanf("%d %d",&a,&b);
  printf("their quotient is %lf", a/b); }
```

(a) 4 bytes (b) 8 bytes
(c) 16 bytes (d) Input dependent

**Answer: (b) Here the calculated double value is not stored anywhere and just printed so it does not consume any space. Thus, resulting in 4+4=8 bytes.**

14. What will be the size of the following program in a 64-bit processor with gcc compiler? (L2)

```
void main()
{ int ;
  printf("hai"); }
```

(a) 4 bytes (b) 0 bytes
(c) 1 bytes (d) 8 bytes

**Answer: (b) Memory is allocated only to variables. So useless data type doesn't consume memory.**

15. What is the output of the program? (L4)

```
void main()
{ printf("%d",int a=10); }
```

(a) 10 (b) Error
(c) 0 (d) Prints nothing

**Answer: (b) Declaration cannot be inside any 'printf' arguments.**

16. What is the output of the code in turbo c 3.5 compiler? (L2)

```
void main()
{ char aabbccddeeffgghhiijjkkllmmnnooppqqrrss= 's';
  char aabbccddeeffgghhiijjkkllmmnnooppqqrrssttuuvv= 'c'
  printf("%c ", aabbccddeeffgghhiijjkkllmmnnooppqqrrss); }
```

(a) Prints nothing (b) Compilation error
(c) c (d) s

**Answer: (b) The variable name is significant only for 32 characters and after that it will not be considered. So here the compiler throws multiple declaration error. It is to be noted that gcc compiler doesn't has that constrain.**

17. What is the ASCII value of '\f'? (L2)

(a) 8 (b) 9 (c) 12 (d) 13

**Answer: (c)**

18. What is the ASCII value of '\t'? (L2)

(a) 9 (b) 12 (c) 11 (d) 45

**Answer: (a)**

19. What is the ASCII value of '\a'? (L2)

(a) 14 (b) 8 (c) 7 (d) 23

**Answer: (c)**

20. What is the ASCII value of '\r'? (L2)

(a) 11 (b) 15 (c) 13 (d) 3

**Answer: (c)**

21. What is the ASCII value of '\b'? (L2)

(a) 14 (b) 8 (c) 7 (d) 5

**Answer: (b)**

22. What is the ASCII value of '\v'? (L2)

(a) 14 (b) 11 (c) 7 (d) 1

**Answer: (b)**

23. What is the ASCII value of '\\'? (L2)

(a) 80 (b) 76 (c) 92 (d) 64

**Answer: (c)**

24. What is the ASCII value of ' \' '? (L2)

(a) 39 (b) 73 (c) 81 (d) 90

**Answer: (a)**

25. What is the output of the code? (L4)

```
void main()
{ int a;
  float b;
  printf("%d %f",b=5.14); }
```

(a) 0 Garbage_value (b) 5 Garbage_value

(c) 5.140000 Garbage_value (d) Garbage_value 5.140000

**Answer: (d) The float value matches itself to the specifier and the undefined specifer takes the garbage value.**

**26.** What is the output of the code? (L3)

```
void main()
{ int a;
  float b;
  printf("%f %d",a=3,b=5.14); }
```

(a) 0 garbage_value
(b) 0 0
(c) 5.140000 3
(d) 3.000000 5

**Answer: (c) In spite of wrong order the arguments matches to the specifiers by themselves.**

**27.** What is the output of the code? (L3)

```
void main()
{ float b=0.23;
  printf("%f %f",b=5.14,b); }
```

(a) 5.140000 5.140000
(b) 5.140000 .230000
(c) Garbage_value Garbage_value
(d) 0.230000 0.230000

**Answer: (b) It just executes from right to left.**

**28.** What is the output of the code? (L2)

```
void main()
{ float g;
printf("%x",g); }
```

(a) Garbage_value
(b) 0
(c) 0x00
(d) Error

**Answer: (a) The initial value 0.000000 is a float and does not have a hexadecimal value. The internal difference in the interpretation of the formats results in some garbage values.**

**29.** What is the output of the code? (L3)

```
void main()
{ int a=3;
if(a==3)
{ switch(a)
  { case 2: printf("case1");
                  int x=1;
                  break;
            case 3: printf("%d",x);
                  x=2; } }
else
{ case 4: printf("%d",x);
                  x=2; } }
```

(a) Case 1
(b) Prints nothing
(c) Error
(d) None of the above

**Answer: (c) Even though the 'else' part is not accessed and 'case 4' is also not accessed, yet it throws the same errors as linking will not be possible.**

**30.** Which of the statements will check "a==0"? (L3)

(a) switch(a){case 0: printf("true");}
(b) (a==0)?printf("true"):printf("false");
(c) Both (a) and (b)
(d) All of the above

**Answer: (d) Option (a) uses switch to match (a) with '0'. Option (b) uses ternary operator (?:) to check it.**

**31.** What should be the statement for the output to be "hai"? (L4)

```
void main()
{ int a=0;
  __________
{ case 1: printf("hai");
              break;
    case 0: printf("hai");} }
```

(a) switch(a==0)
(b) switch(a)
(c) both (a) and (b)
(d) None of the above

**Answer: (c) Here both the 'case' blocks has 'printf' with 'hai'. Option (a) will give switch value '1' and option (b) gives '0'. So both suit.**

**32.** What is the difference between 'switch' and 'if..else' ladder? (L2)

(a) 'switch' can have more number of cases and more efficient.
(b) 'if…else' can work on all data types and it is more efficient.
(c) They both are equal for the compiler.
(d) Both are equally efficient but 'switch' has a lesser execution time and better.

**Answer: (c) For the user there may be any difference in terms of datatype and coding complexities. But for the compiler they both are equal structures just as an alias of one another.**

**33.** What is the output of the code? (L3)

```
void main()
{ int a=0;
  a: printf("a");
      b: printf("b");
      (a++%2)? goto a: goto b; }
```

(a) ababab....(infinitely)
(b) bababa....(infinitely)
(c) error
(d) None of the above

**Answer: (c) Expected expression before 'goto'.**

34. What is the output of the code? (L3)

```
void main()
{ int i=10;
 a:
 if(a%2)
 goto b;
     printf("%d",i);
 i++;
 b: goto a; }
```

(a) 101010...(infinitely)
(b) 10 then the complier traps at the point, execution doesn't stop
(c) 10
(d) None of the above

**Answer: (d) The compiler just traps. '10' is even so the condition satisfies. Continuously it jumps from 'a' to 'b' and 'b' to 'a'. So no progress, and execution doesn't stop.**

35. What is the output of the code? (L3)

```
void main()
{ int i;
 a: for(;i<10;)
 for(;i<12;)
 printf("%d",i++);
 goto a: }
```

(a) Trap
(b) 01234567891011 and traps
(c) Error
(d) None of the above

**Answer: (a) The compiler first completely checks the program and then only runs it. It will trap as the 'goto' repeatedly tries to execute that loop. But it doesn't have initialisation. So cannot be run.**

36. What is the difference in the codes? (L3)

```
void main()
{ int a=10;
 while (a<10)
 { printf("%d",a++);
   a- -;}        }
```

```
void main()
{ int a=10;
 for(;a<10;)
  { printf("%d",a);
       a- -;} }
```

(a) No difference
(b) Both are logically same but structurally different
(c) They are logically different but gives same output
(d) One program is an altered version of the other but still logically same

**Answer: (d) Both are loops and has same steps. Only they have difference in the positions.**

37. What is the output of the code? (assuming address of d[0] as f35f5740c) (L3)

```
void main()
{ int d[2]={2,4},*c[2];
  c[0]=&d;
  printf("%d",*(c[0]); }
```

(a) Compiler dependent
(b) 2
(c) Garbage value
(d) Error

**Answer: (b) Array of pointers can also be created.**

38. What is the output of the code? (L4)

```
void main()
{ char d[2],*c;
  d[0]='a';
  d[1]='b';
  c=&d;
  printf("%c",*c); }
```

(a) Compiler dependent
(b) a
(c) Error
(d) None of the above

**Answer: (a) As the format specifier is '%c' only the first element will be printed.**

39. What is the output of the code? (L3)

```
void main()
{ int *a;
  a[1]=5,a[2]=3;
  printf("%d",a[1]);}
```

(a) 8
(b) 4
(c) Error
(d) None of the above

**Answer: (d) Segmentation fault. Here '*a' is not initiated properly, especially for the first.**

40. What is the output of the code? (L3)

```
void main().
{ char d[2],*c;
```

```
d[0]='a';
d[1]='b';
c=&d[0];
printf("%c",*c); }
```

(a) Compiler dependent (b) a
(c) error (d) None of the above

**Answer: (b) Pointers can also be assigned to an array element.**

**41.** What is the output of the code? (L3)

```
void main()
{ char d[2],*c;
  d[0]='a';
  d[1]='b';
  c=&d[0];
  printf("%c",*c[0]); }
```

(a) Compiler dependent (b) a
(c) error (d) None of the above

**Answer: (c) Invalid type argument of unary '*'.**

**42.** What will be the error thrown by the code? (L4)

```
int sqr(char c);
void main()
{ int a=8,b;
  b=sqr(a);
  printf("%d",b);}
  int sqr(int c)
{ return c*c;}
```

(a) 'c' undeclared (first use in this function)
(b) Conflicting types for 'sqr'
(c) Invalid defisssnition of 'sqr' with argument 'int'
(d) None of the above

**Answer: (b) The declaration will be given more priority. It will be save inside first and then the definition will be mapped to it.**

**43.** What is the output of the code? (L4)

```
void main()
{ int a=8,b;
  b=sqr(a);
  printf("%d",b);}
  int sqr(int c)
```

```
{ return c*c;}
int sqr(char c);
```

(a) 64
(b) Garbage value
(c) Error
(d) None of the above

**Answer: (a) In GCC this doesn't have a problem. And the declaration done at the last is also useless. In some compiler this may give error of missing function 'sqr' as it is not declared before 'main'.**

**44.** What is the output of the code? (L3)

```
void sqr(char c);
void main()
{ int a=8,b;
  b=sqr(a);
  printf("%d",b);}
  int sqr(int c)
{ return c*c;}
```

(a) 64
(b) Garbage value
(c) Error
(d) None of the above

**Answer: (c) The return types between the declaration and definition cannot be changed.**

**45.** What will be the error thrown by the code? (L4)

```
void sqr(char c);
void main()
{ int a=8,b;
  b=sqr(a);
  printf("%d",b);}
  int sqr(int c)
{ return c*c;}
```

(a) Conflicting types for 'sqr' in 'int sqr(int c)'
(b) void value not ignored as it ought to be 'b=sqr(a)'
(c) Conflicting types for 'sqr' in 'void sqr(int c)'
(d) Both '(a)' and '(b)'

**Answer: (d) Always declaration will be of primary reference than definition. As the declaration is of void type '(b)' comes.**

**46.** What is the output of the code? (L3)

```
void sqre(char c);
void main()
{ int a=8,b;
```

```
b=sqr(a);
printf("%d",b);}
int sqr(int c)
{ return c*c;}
```

(a) 64 (b) Garbage value
(c) Error, confliction function name (d) None of the above

**Answer: (a) Here the declaration of function 'sqre' is separate and the function 'sqr' is saved separately. So doesn't conflict. And it is not necessary that all declared function must be defined.**

**47.** What should filled in the blank to find the length of the string? (L4)

```
void main()
{ char a[ ]= "hai";
 printf("%lu",____________);}
```

(a) sizeof a / sizeof *a (b) sizeof a / sizeof *a + 1
(c) sizeof a / sizeof *a - 1 (d) None of the above

**Answer: (c) This is because of the default '\0' inside the string.**

**48.** What is the output of the code? (L4)

```
void main()
{ char a[2],b[5]= "hai";
 strcpy(a,b);
 printf("%s", a);}
```

(a) h (b) hai
(c) Error (d) None of the above

**Answer: (b) As strings will be treated as pointer of char, it will automatically get stored in the continuous memory location while copying the value.**

**49.** What is the output of the code? (L4)

```
void main()
{ char a[5]={ };
 strncpy(a, "hello",2);
 printf("%s", a);}
```

(a) he (b) he followed by some garbage
(c) Error (d) Segmentation Fault

**Answer: (b) 'Strncpy' copies the first 'n' characters from the destination to the source. Similarly, the first two will be copied but also garbage from consecutive memory will be printed. It is also important to know that only GCC has this garbage problem. In other compilers like Dev C++ and Turbo C the answer will be '(a)'.**

**50.** What is the output of the code? (L4)

```
void main()
{ char a[5]= "hello";
  memset(a,'\0',2);
  printf("%s", a);}
```

(a) llo (b) Prints nothing
(c) Error (d) Segmentation Fault

**Answer: (b) This is based on the fact that '%s' will print the value until '\0' is reached. Here '\0' is stored in the first two indexes, hence will print nothing. But it will be stored as "\0\0llo\0" internally.**

**51.** What is the output of the code? (L3)

```
void main()
{ printf("%d",ispunct(63));}
```

(a) Some positive number (b) 0
(c) Error (d) Segmentation fault

**Answer: (a) 63 is the ASCII value of '?'.**

**52.** What is the output of the code? (L4)

```
void main()
{ printf("%d",isgraph(63));}
```

(a) Some positive number (b) 0
(c) Error (d) Segmentation fault

**Answer: (a) 'isgraph' is a combination of 'isalnum' and 'ispunct'. Here the value thrown mostly is 32768.**

**53.** What is the output of the code? (L3)

```
void main()
{ int a;
  { int b=9; }
  printf("%d",b);}
```

(a) 9 (b) Unpredictable behaviour
(c) Error (d) None of the above

**Answer: (c) 'b' undeclared (first use in this function). A variable declared in a block cannot be accessed outside the block. The block where the variable is available is known as the scope.**

**54.** What is the output of the code? (L3)

```
void main()
{ int a=10;
  { int a=9;
    printf("%d",a);} }
```

(a) 9
(b) 10
(c) Error
(d) None of the above

**Answer: (a) In any block the local variable will be given more precedence. In such situations there is another problem that the outer value cannot by accessed inside the block.**

**55.** What is the output of the code? (L4)

```
void main()
{ int a=9;
  { if (a<9) }
     (printf("%d",a)); }
```

(a) 9
(b) Garbage value
(c) Error
(d) None of the above

**Answer: (c) Expected expression before '}' in the 'if' statement. A block having 'if' must be properly terminated. Here '}' marks the end of the block, but is present in the 'if' sub-block. Hence, the error is thrown.**

**56.** What is the output of the code? (L3)

```
void main()
{ int a=9;
  { if (a<9)
    (printf("%d",a));} }
```

(a) 9
(b) Garbage value
(c) Error
(d) None of the above

**Answer: (d) Prints nothing. Inside a block when a variable is accessed, it will search whether it is present in the local scope. If there is no such local variable then it will search in the next higher scope. After all such searches will only throw the 'undeclared error'. So here the 'a' in the 'if' condition will refer to previous 'a' with 9. So the condition fails. So 'if' block is not executed. Thus, prints nothing.**

**57.** What will be the output of the code? (L4)

```
void main(int a)
{ printf("%d",a); }
```

(a) 0
(b) Garbage value
(c) Error
(d) None of the above

**Answer: (d) Prints 1. This is because of command line argument.**

**58.** What is the output of the code? (L3)

```
int a=9;
void main()
```

```
{ int a=10;
 { printf(" %d",a);} }
```

(a) 10
(b) 9
(c) Error
(d) Unpredictable behaviour

**Answer: (a) The searching if variable will always be in the order of block pr local scope, function scope and then global scope.**

**59.** What is the output of the code? (L3)

```
int a;
void main()
{ printf(" %d",a); }
```

(a) 0
(b) Error
(c) Unpredictable behaviour
(d) Compiler dependent

**Answer: (a) Global variables when declared will be automatically initialised to default value. It will be 0 or its equivalent. But for local variable, the rule doesn't apply. In GCC it will be initialised to 0, whereas in other compilers has all the possibilities of having garbage values.**

**60.** What is the output of the code? (L3)

```
int a=9;
void main()
{ int a=10;
 printf(" %d",::a); }
```

(a) 9
(b) Error
(c) Unpredictable behaviour
(d) Compiler dependent

**Answer: (b) Expected expression before ':' in '::a' (first colon).**

**61.** What is the output of the code? (L4)

```
int a;
a=10.23;
void main()
{ printf("%d",a); }
```

(a) 0
(b) 10
(c) Error
(d) Unpredictable behaviour

**Answer: (b) It doesn't work normally. Usually, this will give '0'. But for global variable before it is stored in it will be converted to the destination type. So there is no issue with the format specifier and printed the integer value. But the same above warning alone will be thrown.**

**62.** What is the output of the code? (L4)

```
a= 9;
void main()
{ printf("%d",a); }
```

(a) 9
(b) Trap
(c) Error
(d) Unpredictable behaviour

**Answer: (a) Just the same warning will be given. It is to be understood that when a variable is defined in global scope, it gets declared in any required format based on the value stored.**

**63.** What is the output of the code? (L4)

```
        main ( )
{ short int *a,b;
        b= 32768;
        a=&b;
        printf( "%d",*a); }
```

(a) -32768
(b) Segmentation fault
(c) error
(d) None of the above

**Answer: (a) When the range is crossed the values will be converted into negative. These conversions are based on the internal storage. The MSB of the binary representation is used as sign bit. When the value increases, the MSB gets 1 this while converting back into int gives negative values.**

**64.** What is the output of the code? (L4)

```
        main ( )
{ short int *a,b;
        b= 65530;
        a=&b;
        printf( "%d",*a); }
```

(a) -6
(b) 0
(c) Segmentation fault
(d) error

**Answer: (a) The calculation is in such a way that it will be 0 for 32767, when goes further it starts incrementing from -32768. It reaches 0 again on 65536 which is the double of the positive maximum plus 2. Again this increases.**

**65.** What is the output of the code? (L4)

```
        main ( )
{ int a=65530;
        short int a= 35567;
        printf( "%d",a); }
```

(a) 35567
(b) Unpredictable behaviour
(c) Segmentation fault
(d) Error

**Answer: (c) Conflicting types for 'a'.**

**66.** What is the output of the code? (L4)

```
main ( )
{ (short)int a=65530;
printf( "%d",a); }
```

(a) –6
(b) Segmentation fault
(c) Error
(d) None of the above

**Answer: (c) "Expected expression before 'int' ". And because of this " 'a' undeclared (first use in this function)"**

**67.** What is the output of the code? (L4)

```
main ( )
{ int short a=65530;
printf( "%d",a); }
```

(a) –6
(b) Segmentation fault
(c) Error
(d) None of the above

**Answer: (a) The type and modifier can be interchanged.**

**68.** What is the output of the code? (L4)

```
main ( )
{ int (short) a=65530;
printf( "%d",a); }
```

(a) –6
(b) Segmentation fault
(c) Error
(d) None of the above

**Answer: (c) "Expected expression or '(' before 'short' ". And because of this " 'a' undeclared (first use in this function)"**

**69.** What is the output of the code? (L4)

```
main ( )
{ long short int a=4777;
printf( "%d",a); }
```

(a) 4777
(b) Segmentation fault
(c) Error
(d) None of the above

**Answer: (c) Both 'long' and 'short' in declaration specifiers.**

**70.** What is the output of the code? (L4)

```
main ( )
{ short int a=4777;
printf( "%li",a); }
```

(a) 4777
(b) Segmentation fault
(c) Error
(d) None of the above

**Answer: (a) This will just throw a warning for the mismatching format specifiers, but will print normally.**

**71.** What is the output of the code? (L4)

```
main ( )
{ short a=4777;
printf( "%d",a); }
```

(a) 4777
(b) Segmentation fault
(c) Error
(d) None of the above

**Answer: (a) A variable can be declared just by a modifier without the data type.**

**72.** What is the output of the code? (L4)

```
main ( )
{ long float a=3.65e+38;
printf( "%d",a); }
```

(a) 4777596
(b) Segmentation fault
(c) Error
(d) None of the above

**Answer: (c) Both 'long' and 'float' in declaration. But this is allowed in other compilers. 'long' modifier is meant to increase the size of 'float' to 8 bytes. So the range will be -9.22e18 to 9.22e18. The range of normal float is -3.4e38 to 3.4e38. While double has the range of -1.7e308 to 1.7e308. Thus, 'long float' is different from 'double'. 'long float' stores larger values, while 'double' stores longer precision. And "%lf" is the appropriated format specifier.**

**73.** What is the output of the code? (L3)

```
int a=9;
void fun( );
void main ( )
{ extern int a;
  a= 10;
  fun( );
  printf("%d ",a); }
extern void fun( )
{ printf("%d ",a); }
```

(a) 10 10
(b) Unpredictable behaviour
(c) Error
(d) None of the above

**Answer: (a) The additional keyword is just skipped.**

**74.** What is the output of the code? (L3)

```
void main ( )
{ int *b;
  double d[ ]={23.33,32.33};
  b= &d;
  printf("%f %f",b[0],b[1]); }
```

(a) 23 32
(b) Garbage values
(c) Error
(d) None of the above

**Answer: (d) 0.00000 0.000000. Even now the stored value is in some garbage form of integer. On further converting to '%f' in printf, gives 0.000000.**

**75.** What will be the error thrown by the code? (L4)

```
void main ( )
{ void *b;
  double d[ ]={23.33,32.33};
  b= &d;
  printf("%f %f",(float)b[0],(float)b[1]); }
```

(a) Invalid use of void variable 'b'
(b) Void value not ignored as it ought to be
(c) No error
(d) None of the above

**Answer: (d) Invalid use of void expression.**

**76.** What is the output of the code? (L4)

```
void main ( )
{ int a[2],*b;
  b=&a;
  double d[]={23.33,32.33},*c;
  c=&d;
  b= c;
  printf("%d %d",b[0],b[1]); }
```

(a) 23 32
(b) Garbage values
(c) Error
(d) None of the above

**Answer: (b) This is not a proper conversion. While assigning pointers implicit conversion cannot work on the values leaving the address.**

**77.** What is the output of the code? (L3)

```
void main ( )
{ int a = (int) 4.56;
  printf("%d ",a); }
```

(a) 4
(b) Error
(c) Unpredictable behaviour
(d) None of the above

**Answer: (a) This is explicit conversion. This is manually changing the data type of the constant before storing it. Here even if '(int)' is not given it will be internally converted. That is implicit conversion. The type specified as '(int)' is also called cast operator.**

**78.** What is the output of the code? (L3)

```
void main ( )
{ int a = (int) 4;
  printf("%d ",a); }
```

(a) 4
(b) Error
(c) Unpredictable behaviour
(d) None of the above

**Answer: (a) Here '4' is an integer constant. So 'int' is getting converted into 'int'. But practically nothing happens inside the compiler.**

**79.** What is the output of the code? (L3)

```
void main ( )
{ int a[2];
  float b[2] = {98.99,95.65};
  a[0] = (int) b[0];
  a[1] = (int) b[1];
  printf("%d %d",a[1],a[2]); }
```

(a) 98 95
(b) Unpredictable behaviour
(c) Error
(d) None of the above

**Answer: (a) This is the only way in which an array can be converted into another type.**

**80.** What is the output of the code? (L3)

```
void main()
{ char *s;
  s = (char *) malloc(15);
  strcpy(s, "hello");
  printf("%s",s); }
```

(a) Strucirhello
(b) Garbage symbols
(c) Error
(d) Segmentation fault

**Answer: (a) 'malloc' creates a memory location of specified size and returns a pointer to the location. The size is the argument given to it. It will be a void pointer amd can be converted to any data type. It requires the header file 'stdlib.h'.**

81. What is the output of the code? (L3)

```
void main()
{ char *s;
  s = malloc(15);
  strcpy(s, "hello");
  printf("%s",s); }
```

(a) hello
(b) Unpredictable
(c) Error
(d) Segmentation fault

**Answer: (a) Even if the conversion is not present, it will automatically converts the type while assigning.**

82. What is the output of the code? (L3)

```
void main()
{ char *s;
  s = malloc(15);
  printf("%s",s); }
```

(a) Prints nothing
(b) Error
(c) Segmentation fault
(d) Compiler dependent

**Answer: (d) In GCC it will be initialised to zero so prints nothing. But the functionality of the function doesn't involve initialising the memory to null. So in other compilers like Turbo C it may return garbage value.**

83. What is the output of the code? (L4)

```
void main()
{ char *s;
  s = malloc(3);
  strcpy(s, "hello");
  printf("%s",s); }
```

(a) hello
(b) Error
(c) Segmentation fault
(d) Unpredictable

**Answer: (d) It may be 'a' or 'c'. Mostly in GCC the output will be 'a'. Here the allocated memory is small than the copied string. But as strings are continuous memory, it is copied fully using the nearby memory location. Sometimes the neighbouring location may be unavailable and that will cause the segmentation fault.**

84. What is the output of the code? (L3)

```
void main()
{ char *s;
  s = (char *) malloc(20);
  strcpy(str, "hello_world!");
  printf(" %s",s); }
```

(a) hello_world! (b) hello_wo
(c) Error (d) None of the above

**Answer: (a) Strings are always stored in continuous memory locations. So will be printed till '\0' will.**

85. What is the output of the code? (L4)

```
void main()
{ int *s;
  s = (int *) malloc(0);
  printf(" %lu",sizeof(s)); }
```

(a) 8 (b) 0
(c) Error (d) None of the above

**Answer: (a) Every allocated variable will be associated to a register. Even if it is of size 0, it will give 8 only.**

86. What is the output of the code? (L3)

```
void main()
{ int *s;
  s = (int *) malloc(0);
  s[1]=4;
  printf(" %lu",sizeof(s)); }
```

(a) 4 (b) 0
(c) Error (d) None of the above

**Answer: (d) 8 will be the output as usual. Even though only one variable is stored, it has the whole register and gives 8.**

87. What is the output of the code? (L3)

```
void main()
{ int *s;
  s = (void *) malloc(0);
  s[1]=4;
  printf(" %d", s[1]); }
```

(a) 4 (b) Unpredictable behaviour
(c) error (d) None of the above

**Answer: (a) (void *) takes the data type of the left variable. This works exactly as declaring without any data type. Here the (void *) is useless.**

88. What is the output of the code? (L4)

```
void main()
{ int *s;
  s = (int *) malloc(3+ 'a');
```

```
s[1]=3;
printf("%lu", sizeof(s)); }
```

(a) 8
(b) Unpredictable behaviour
(c) Error
(d) None of the above

**Answer: (a) 3+ 'a' is just computed to 100.**

**89.** What is the output of the code? (L3)

```
void main()
{int *s;
 s = (int *) malloc(10000000000);
 printf("%lu", sizeof(s)); }
```

(a) 8
(b) Unpredictable behaviour
(c) Error
(d) None of the above

**Answer: (a) Usual behaviour. 'malloc' can allocate the maximum size of the physical memory. The size specified can be the RAM size.**

**90.** What will the second printf statement give? (assume input given to 'n' is 10) (L4)

```
void main()
{int n;
 char *s;
 printf("enter size");
 scanf("%d",&n);
 s = (char *) malloc(n);
 memcpy(s, "hello",10);
 printf("%d", s); }
```

(a) hello
(b) Unpredictable behaviour
(c) Error
(d) Segmentation fault

**Answer: (a) malloc can also be called with user specified variable.**

**91.** What will be the output of the code? (L3)

```
void main()
{char *s;
s=(char *) malloc(15);
free(s);
free(s);
printf("\n%s",s);}
```

(a) Prints nothing
(b) Error
(c) Segmentation fault
(d) None of the above

**Answer: (d) Prints "*** Error in `./a.out`: double free or corruption (fasttop): pointer address *** Aborted (core dumped)". This is due to freeing the memory which is already free. Such errors are called runtime errors, they mostly occur in such situations of dynamic sizing**

**92.** What will be the output of the code? (L3)

```
void main()
{char *s;
s=(char *) malloc(15);
strcpy(s, "hai");
free(s);
strcpy(s, "hello");
printf("\n%s",s);}
```

(a) hello (b) Error
(c) Segmentation fault (d) Compiler dependent

**Answer: (a) Value can be stored or allocated again even after freeing the memory.**

**93.** What will be the output of the code? (L3)

```
void main()
{char *s;
s=(char *) malloc(15);
strcpy(s, "hai");
free(s);
strcpy(s, "hello");
printf("\n%lu",sizeof(s));}
```

(a) 8 (b) 6
(c) Error (d) Segmentation fault

**Answer: (a) Once the pointer is allocated a memory, freeing it will clear it all, but still have the same register and so size is the same 8 bytes.**

**94.** What is the output of the code? (L3)

```
struct s
{ int a;
}b,c;
void main ( )
{ b.a = 4;
  c.a=4;
printf("%d %d",b.a,c.a); }
```

(a) 4 4 (b) 0 0
(c) Error (d) Unpredictable behaviour

**Answer: (a) Here 'b' and 'c' are multiple instances for the struct.**

95. What is the output of the code? (L3)

```
struct s
{ int a; };
void main ( )
{ struct s b,c;
  b.a = 4;
  c.a=4;
printf("%d %d",b.a,c.a); }
```

(a) 4 4
(b) 0 0
(c) Error
(d) Unpredictable behaviour

**Answer: (a) This is another method of creating instances to a structure.**

96. What is the output of the code? (L3)

```
struct s
{ int a; };
void main ( )
{ struct s b;
  printf("%d ",b.a); }
```

(a) 0
(b) Error
(c) Unpredictable behaviour
(d) Compiler dependent

**Answer: (c) Struct variables are not initialised with 0. Unless they are accessed they rest in some random memory with some garbage value.**

97. What is the output of the code? (L3)

```
struct s
{ int a; };
void main ( )
{ s b;
  b.a = 4;
  printf("%d ",b.a); }
```

(a) 4
(b) 0
(c) Error
(d) Unpredictable behaviour

**Answer: (c) Struct is a user defined data type. So declaring an instance without struct keyword makes it unknown to the compiler.**

98. What is the output of the code? (L3)

```
struct s
{ int a; };
void main ( )
```

```
{ struct s b;
  b.a = 4;
  printf("%lu ",sizeof(b)); }
```

(a) 4
(b) 8
(c) Error
(d) Compiler dependent

**Answer: (a) The struct has only one variable of size four. So the size of the structure instance is the same, 4 bytes.**

**99.** What is the output of the code? (L4)

```
struct s
{ int a;
  char c; };
void main ( )
{ struct s b;
  b.a = 4;
  b.c= 'd';
  printf("%lu ",sizeof(b)); }
```

(a) 4
(b) 8
(c) 5
(d) Compiler dependent

**Answer: (d) In GCC this gives 8. In other compilers it may give 5. This depends also on the architecture of the compiler. Here in GCC it gets padded automatically. For any variable the size of the largest variable is allocated, even if it needs less memory. The adding of additional memory to make up that length is called padding. It basically tries to allocate a register as a whole. Which is 8 bytes.**

**100.** What is the output of the code? (L4)

```
struct s
{ int a; };
struct e
{ int a; };
void main ( )
{ struct s b;
  struct e v,*c;
  b.a=4;
  v.a=9;
  c=&b;
  printf("%d %d", (*c).a,v.a); }
```

(a) 4 9
(b) Error
(c) Unpredictable behaviour
(d) Segmentation fault

**Answer: (a) Here the pointer automatically gets converted from one structure type to another.**

**101.** What is the output of the code? (L4)

```
struct s
{ int a; };
struct e
{ int a;
  float vb; };
void main ( )
{ struct s b;
  struct e v,*c;
  b.a=4;
  v.vb=15.23;
  c=&b;
  printf("%d %f", (*c).a,v.vb); }
```

(a) 4 15.230000 (b) Error
(c) Unpredictable behaviour (d) Compiler dependent

**Answer: (a) Even when the structures are completely different and have different size, the same conversion is maintained.**

**102.** What is the output of the code? (L4)

```
struct s
{ int a; };
struct e
{ int a;
  float vb; };
void main ( )
{ struct s b;
  struct e v,*c;
  b.a=4;
  v.vb=15.23;
  c=&b;
  printf("%d %f", (struct s)(*c).a,v.vb); }
```

(a) 4 15.230000 (b) Error
(c) Unpredictable behaviour (d) Compiler dependent

**Answer: (b) Conversion to non-scalar type requested. Explicit conversion to struct type is not possible.**

**103.** What is the output of the code? (L4)

```
struct s
{ int a; };
struct e
{ float a; };
void main ( )
{ struct s b,*c;
  struct e v;
  b.a=4;
  v.a=4.2;
  c=&v;
  printf("%d %f", b.a,(*c).a); }
```

(a) 4 4.200000
(b) Error
(c) Unpredictable behaviour
(d) None of the above

**Answer: (d) 4 0.000000. The variable with same name is of type 'int' in the struct 'b'. After converting the variable in the converted struct is of type 'float'. So compiler expects '%d' after conversion for the common variable. On trying to print float with integer gives 0.000000.**

**104.** What is the output of the code? (L4)

```
struct s
{ int a; };
struct e
{ float a; };
void main ( )
{ struct s b,*c;
  struct e v,*d;
  b.a=4;
  v.a=4.2;
  d=&v;
  c=d;
  printf("%d %f", b.a,(*c).a); }
```

(a) 4 0.000000
(b) Error
(c) Unpredictable behaviour
(d) None of the above

**Answer: (a) Even the conversion is between pointers of different structures, there is no proper or complete implicit conversion taking place. The same as previous case happens.**

105. What is the output of the code? (L3)

```
struct s
{ int a; };
void fun(struct s f)
{ printf("%d",++f.a); }
void main ( )
{ struct s b;
  b.a=4;
  fun(b);
  printf("%d", b.a); }
```

(a) 54
(b) 55
(c) Error
(d) Compiler dependent

**Answer: (a) This works as any normal variable passed as argument to a user defined function.**

106. What is the output of the code? (L3)

```
struct e
{ int a;
  float m;
  auto char c;};
void main ( )
{ struct e s1;
  s1.a=4;
  s1.m=56.25;
  s1.c= 'f';
  printf("%f ", s1.m); }
```

(a) 56.250000
(b) Error
(c) Compiler dependent
(d) Segmentation fault

**Answer: (b) This behaves in the same way as 'static' modifier. This is not a legal way of declaring structure variables. It is clear that modifiers cannot be used in this way.**

107. What is the output of the code? (L3)

```
struct e
{ int a; };
struct s
{ int *h; };
void main ( )
{ struct e s1;
  s1.a=4;
  struct s s2;
```

```
s2.h=&(s1.a);
printf("%d ", *(s2.h)); }
```

(a) 4
(b) Error
(c) Segmentation fault
(d) Unpredictable behaviour

**Answer: (a) This works as a normal pointer.**

**108.** What is the output of the code? (L3)

```
typedef int c;
void main()
{ c a;
  a=12;
  printf("%d",a); }
```

(a) 12
(b) Error
(c) Unpredictable behaviour
(d) None of the above

**Answer: (a) Typedef defines a new datatype. It simply associates, a user defined name with some other data type. Say here, 'c' can be used instead of int throughout the program.**

**109.** What is the output of the code? (L3)

```
typedef float c;
void main()
{ c a[2];
  a[1]=3.22;
  a[0]=5.22;
  printf("%f %f",a[1],a[0]); }
```

(a) 3.220000 5.220000
(b) 5.220000 5.220000
(c) Garbage_value 5.220000
(d) None of the above

**Answer: (a) This works normally as a float array.**

**110.** What is the output of the code? (L3)

```
void main()
{ typedef float c;
  c a[2];
  a[1]=3.22;
  a[0]=5.22;
  printf("%f %f",a[1],a[0]); }
```

(a) 3.220000 5.220000
(b) 5.220000 5.220000
(c) Garbage_value 5.220000
(d) None of the above

**Answer: (a) Type definition may be confined to a particular block and will work normally.**

**111.** What is the output of the code? (L3)

```
void main()
{ typedef {
  float g;
  int h;
  char r; }
  g =56.22;
  h=3;
  r= 's';
  printf("%f %d %c",g,h,s); }
```

(a) 56.220000 3 s
(b) 56.220000 Garbagge_values
(c) Error
(d) Compiler dependent

**Answer: (c) Type definition is not valid here. Multiple types cannot be defined like this.**

**112.** What is the output the code? (L2)

```
void main()
{ typedef float f;
  typedef f m;
  m g =56.22;
  printf("%f",g); }
```

(a) 56.220000
(b) 56.219999
(c) Unpredictable behaviour
(d) None of the above

**Answer: (a) It works normally.**

**113.** What is the output of the code? (L2)

```
void main()
{ typedef float f;
  f =56.22;
  printf("%f",f); }
```

(a) 56.220000
(b) 56.219999
(c) Unpredictable behaviour
(d) None of the above

**Answer: (d) It throws error. After type definition, the type is directly assigned a value. This is not legal.**

**114.** What is the output of the code? (L3)

```
void main()
{ typedef int m;
  typedef int n;
  n f =56;
  printf("%d",f); }
```

(a) 56
(b) Unpredictable behaviour
(c) Error
(d) Compiler dependent

**Answer: (a) Multiple types can be defined with same base type.**

**115.** What is the output of the code? (L3)

```
void main()
{ typedef int m;
  typedef int n;
  n f =56;
  m e=65;
  printf("%d %d",f,e); }
```

(a) 56 65
(b) 56 Garbage_Value
(c) Error
(d) Compiler dependent

**Answer: (a) This too works normally.**

**116.** What is the output of the code? (L3)

```
typedef union s v;
typedef void * g;
union s
{ int f; };
void main()
{  v l;
   g m=&l;
   l.f=15;
   printf("%d",(*(v *)m).l); }
```

(a) 15
(b) Garbage value
(c) Error
(d) Segmentation fault

**Answer: (a) This works normally.**

**117.** What is the output of the code? (L2)

```
void main()
{ enum numbers { zero, one, two };
  enum numbers a=two;
  printf("%d",a); }
```

(a) 2
(b) Error
(c) Unpredictable behaviour
(d) None of the above

**Answer: (a) Enum is a set of variables. The variables gets names starting from zero for the first variable. So when an integer is assigned with a variable. It gets the corresponding integer value.**

118. What is the output of the code? (L2)

```
void main()
{ enum numbers { zero=9, one, two };
  enum numbers a=two;
  printf("%d",a); }
```

(a) 11 (b) 2
(c) Error (d) None of the above

**Answer: (a) When the first variable is initialised to a value, it gets the consecutive values. This is allowed instead to taking the default '0' for first value.**

119. What is the output of the code? (L3)

```
void main()
{ enum numbers { zero, one=9, two };
  enum numbers a=two;
  printf("%d",a); }
```

(a) 10 (b) 2
(c) Error (d) None of the above

**Answer: (a) Value can also be initialised from any required spot. Then the consecutive variables gets the consecutive values.**

120. What is the output of the code? (L3)

```
void main()
{ enum numbers { zero=5, one=2, two=9, three };
  enum numbers a=one;
  enum numbers b=three;
  printf("%d %d",a,b); }
```

(a) 2 10 (b) Unpredictable behaviour
(c) Error (d) None of the above

**Answer: (a) Even this is allowed. The values initialised need not be in order.**

121. What is the output of the code? (L3)

```
void main()
{ enum numbers { zero=5, one= -2, two=9, three };
  enum numbers a=one;
  enum numbers b=three;
  printf("%d %d",a,b); }
```

(a) –2 10 (b) Unpredictable behaviour
(c) Error (d) None of the above

**Answer: (a) Even negative values are allowed.**

**122.** What is the output of the code? (L3)

```
void main()
{ enum vehicle { car, bus='s', bike, lorry };
  enum vehicle a=bike;
  printf("%c",a); }
```

(a) u
(b) Unpredictable behaviour
(c) Error
(d) Compiler dependent

**Answer: (a) This works because internally 'int' and 'char' has the same architecture.**

**123.** What is the output of the code? (L3)

```
void main()
{ enum vehicle { car, bus=12, bike, lorry } ;
  typedef enum vehicle k;
  k a=bike;
  printf("%d",a); }
```

(a) 13
(b) Unpredictable behaviour
(c) Error
(d) Compiler dependent

**Answer: (a) Typedef is also possible with enumerator.**

**124.** What is the output of the code? (L4)

```
void main()
{ typedef enum vehicle { car, bus=12, bike, lorry } k;
  k a=bike;
  printf("%d",a); }
```

(a) 13
(b) Unpredictable behaviour
(c) Error
(d) Compiler dependent

**Answer: (a) This is a method of declaring the enumerated type with a new type. Enum along with typedef, makes the code very simple and readable.**

**125.** What will be the error thrown by the code? (L3)

```
void main()
{ enum vehicle { car, bus, car}f;
  f=car;
  printf("%d",f); }
```

(a) Redeclaration of variable 'car'
(b) Redeclaration of enumerator 'car'
(c) 'car' redeclared
(d) No error

**Answer: (d) The identifiers declared as a part of the enum are referred as enumerator.**

**126.** What is the output of the code? (L4)

```
void main()
{ enum vehicle { car, bus=9}f;
  f=bus;
  f=12;
  printf("%d",f); }
```

(a) 12 (b) 9
(c) Error (d) None of the above

**Answer: (a) Enum variables may get any integer constant.**

**127.** What is the output of the code? (L4)

```
void main()
{ enum vehicle { car, bus=9}f;
  f=bus;
  float w=(float)f;
  printf("%f",w); }
```

(a) 9.000000 (b) 0.000000
(c) Error (d) Segmentation fault

**Answer: (a) It works as a normal type cast for int to float.**

**128.** What is the output of the code when "hai.txt" is already created? (L2)

```
void main()
{ FILE *f;
  f=fopen("hai.txt", "w");
  fprintf(f, "hai");
  printf("hai printed");
  fclose(f); }
```

(a) hai printed (b) Unpredictable behaviour
(c) Error (d) None of the above

**Answer: (a) This is file handling in 'C'. FILE is used to create a file pointer. 'fopen' opens the file in the specified format. "w" represents write mode. 'fprintf' and 'fcscanf' are used to print into and scan from files. Also the file pointer is important as a parameter. 'fclose' is used to close the file. It is also to be noted that the file should be in the same folder or will be created in the same folder, unless the address is specified.**

**129.** What is the output of the code when "hai.txt" is not created already? (L3)

```
void main()
{ FILE *f;
  f=fopen("hai.txt", "w");
```

```
fprintf(f, "hai");
printf("hai printed");
fclose(f); }
```

(a) hai printed
(b) Unpredictable behaviour
(c) Error
(d) None of the above

**Answer: (a) While opening in write mode if the file is not present already. It will automatically create it.**

130. What is the output of the code when "hai.txt" is the same file used in above question? (L2)

```
void main()
{ FILE *f;
  f=fopen("hai.txt", "r");
  char s[10];
  fscanf(f, "%s",s);
  printf("%s",s);
  fclose(f); }
```

(a) hai
(b) Computational garbage
(c) Error
(d) None of the above

**Answer: (a) This opens the file and reads the first 10 characters into 's'. As the content is less than 10 characters, 'hai' is read and the same is printed on the console.**

131. What is the output of the code when "i.txt" has "hai::bye"? (L4)

```
void main()
{ FILE *f;
  char *s,*c;
  f=fopen("i.txt", "r");
  fseek(f,0,SEEK_END);
  long fs=ftell(f);
  s=(char *) malloc(fs);
  rewind(f);
  c=strtok(b, "::");
  while(c !=NULL)
{ printf("%s\t",c);
  c=strtrok(NULL, "::"); } }
```

(a) hai::bye
(b) hai
(c) hai bye
(d) None of the above

**Answer: (c) 'strtok' is a special function used for converting a part of a string into another. This serves as setting delimiter in files. This is an efficient way of adding delimiter into a file.**

132. What is the output of the code when "l.txt" does not exists? (L4)

```
void strerror(int i)
{ printf("Success"); }
 void main()
{ FILE *f;
 f=fopen ("l.txt", "r");
 if(f==NULL)
 strerror(errno); }
```

(a) Success
(b) File or directory does not exist
(c) The code has error
(d) None of the above

**Answer: (c) The redefinition of 'strerror' doesn't match with the internal definition. Overwriting of an internal function must follow the same syntax.**

133. What is the output of the code when "l.txt" does not exists? (L4)

```
char * strerror(int i)
{ return "Success"; }
void main()
{ FILE *f;
 f=fopen ("l.txt", "r");
 if(f==NULL)
 { char *b=strerror(errno);
  printf("%s",b); } }
```

(a) Success
(b) File or directory does not exist
(c) Segmentation fault
(d) The program has error

**Answer: (a) This perfectly overwrites the internal function. Now any error will be printed as "Success" inside the program.**

134. What is the output of the code when "l.txt" does not exists? (L4)

```
void perror(char *v)
{ printf("Success"); }
void main()
{ FILE *f;
 f=fopen ("l.txt", "r");
 if(f==NULL)
 perror("Error"); }
```

(a) Success
(b) Error success
(c) Segmentation fault
(d) The program has error

**Answer: (d) The definition of 'perror' has an argument 'const char *'. So this causes the problem. The error will be "conflicting types for 'perror' ".**

**135.** What is the output of the code? (L3)

```
# define l 9
void main()
{ printf("%d",l);}
```

(a) 9
(b) Error
(c) Garbage symbol
(d) None of the above

**Answer: (a) # define is called macro. It defines a program scope variable 'l'. Wherever 'l' is found, it will be replaced with 9. '#' is pre-processor directive, which directs the pre-processor to do the specified instruction.**

**136.** What will be the error thrown by the code? (L4)

```
# define 5 9
void main()
{ printf("%d",5);}
```

(a) Invalid identifier '5'
(b) Invalid macro
(c) No error
(d) None of the above

**Answer: (d) Macro names must be identifiers. '5' is not a valid identifier. It is also to be noted that unlike usual errors, macros show the line number and function context of the error. The same will be followed for undefined function calls.**

**137.** What is the output of the code? (L4)

```
# define int float
void main()
{ int d=9.9;
  printf("%f",d);}
```

(a) 9.900000
(b) 9.000000
(c) Error
(d) None of the above

**Answer: (a) The macro name doesn't cause any problem. It also allows to change datatypes. In the whole program, 'int' is treated as 'float'.**

**138.** What is the output of the code when "hai.txt" exists? (L3)

```
# include "hai.txt"
void main()
{ printf("hai"); }
```

(a) hai
(b) Error, "#include doesn't opens text files"
(c) Compiler traps
(d) No output

**Answer: (a) Program can also include text files. But it cannot be used in anyway.**

**139. What is the output of the code? (statements for compilation: "gcc filename.c", "./a.out 5") (L4)**

```
void main(int argc, char *argv[ ])
{ printf("%d %s",argc,argv[1]); }
```

(a) 2 5
(b) 1 5
(c) Error
(d) Segmentation fault

**Answer: (a) This is the way of getting "command line inputs". 'argc' gets default value, the number of argument. While 'argv' is the argument vector. It is an array of strings. It is also to be noted that the 'argc' has the count of number of strings in the command for execution. It also counts './a.out'.**

**140.** What is the output of the code? (statements for compilation: "gcc filename.c", "./a.out 5") (L4)

```
void main(int argc, char *argv[ ])
{ printf("%s",argv[0]); }
```

(a) (null)
(b) ./a.out
(c) Error
(d) Segmentation fault

**Answer: (b) Like 'argc', the vector 'argv' also takes './a.out' inside.**

**141.** What is the output of the code? (statements for compilation: "gcc filename.c", "./a.out 5") (L4)

```
void main(int argc, char *argv[ ])
{ printf("%s",argv[2]); }
```

(a) (null)
(b) ./a.out
(c) Error
(d) Segmentation fault

**Answer: (a) Vectors values not initialised has (null) by default.**

**142.** What is the output of the code? (statements for compilation: "gcc filename.c", "./a.out 5") (L4)

```
void main(int f, char **d)
{ if(d[1][0]== '5')
  printf("%s",d[1][0]); }
```

(a) 5
(b) Garbage symbol
(c) Error
(d) Segmentation fault

**Answer: (a) Here the argument is treated as a number and checks it similar to a number.**

**143.** What is the output of the code? (statements for compilation: "gcc filename.c", "./a.out 5.35") (L4)

```
void main(float f)
{ printf("%f",f); }
```

(a) 5.350000
(b) 5.000000
(c) 0.000000
(d) The code has errors, cannot be executed

**Answer: (c) As usual, the values go useless. 'f' gest default value.**

**144.** What is the output of the code? (statements for compilation: "gcc filename.c", "./a.out 5.35") (L4)

```
void main(void *f[ ])
{ printf("%s",(char **)f[1][0]); }
```

(a) 5.35
(b) Segmentation fault
(c) Compiler traps
(d) The code has errors, cannot be executed

**Answer: (d) "invalid use of void expression".**

**145.** What is the output of the code? (statements for compilation: "gcc filename.c", "./a.out 5.35") (L4)

```
void main(void *f[ ])
{ char *s=*(char *)f[1];
  printf("%s",s); }
```

(a) 5.35
(b) Segmentation fault
(c) Compiler traps
(d) The code has errors, cannot be executed

**Answer: (b) The usage of pointers has the issue while converting from void pointer to char pointer.**

**146.** What is the output of the code? (statements for compilation: "gcc filename.c", "./a.out 5.35") (L4)

```
void main(void *f[ ])
{ char **s=(char **)f;
  printf("%s",s[1]); }
```

(a) 5.35
(b) Segmentation fault
(c) Compiler traps
(d) The code has errors, cannot be executed

**Answer: (b) Even now there is an issue in the address.**

**147.** What is the output of the code? (L4)

```
main()
{ int a=9;
 printf("%#o ",a); }
```

(a) 9 (b) Error
(c) 11 (d) 011

**Answer: (d) '#' in the format specifier is used to print the number along with the notation of representation. It is to be noted that octal numbers are represented with a '0' in the beginning.**

**148.** What is the output of the code? (L4)

```
main()
{ int a=15;
 printf("%#x ",a); }
```

(a) 0f (b) Error
(c) f (d) 0xf

**Answer: (d) '#' in the format specifier is used to print the number along with the notation of representation. It is to be noted that hex numbers are represented with a '0x' in the beginning.**

**149.** What is the output of the code? (L4)

```
void main()
{ int a=9,b=5,c=9,d=c<<a|b;
 printf("%d ",d); }
```

(a) 16384 (b) 4613
(c) 117 (d) Error

**Answer: (b) a|b gives 13 and then 'c' is shifted 13 digits to the left.**

**150.** What is the output of the code? (L4)

```
void main()
{ char c[ ]= {"h"[2]="i"};
 printf("%s",c ); }
```

(a) h (b) Compilation error
(c) Logical error (d) None of the above.

**Answer: (c) while storing the compiler tries to jump from one memory location to another and results in segmentation fault.**

# 6 C++

In most of the interviews questions in C++ will be theoretical and oral. So this part will be helpful in making you strong in C++ concepts.

1. What are tokens? How many are there in C++?

   The smallest element of the program that is reasonably correlated by collection of characters forms a token. There are 5 types of tokens.

   - Keywords
   - Identifiers
   - Literals
   - Punctuators
   - Operators

2. How is type definition done?

   The typedef is a keyword that informs the complier to redefine the existing data types in C++ program to their own terms of data types without any meaning change.

   Syntax: typedef exisiting datatype newdatatype

   Example: typedef int Distances

3. What is user defined datatype? How is it useful?

   User-define data types are:

   - Structure
   - Classes
   - Typedef
   - Enumeration

   *Structure*: To collect the data of various other data types and process together as one single entity. It starts with keyword called Struct and followed by user defined type

   *Classes*: The purpose of this data type is for reusability, the program can have its own class with Class Keyword. The class consists of data members and member function enclosed within parenthesis {}

   *Typedef*: The typedef is a keyword that informs the complier to redefine the existing data types in C++ program to their own terms of data types without any meaning change.

*Enumeration*: This data type provides the integer constant by specifying in terms of identifiers or variables (names), thus the integer constant referred by variables is called enumeration. By default it starts at zero (0). The Keyword used is enum.

**4.** For a program with no object oriented concept involved, which language should be preferred, C or C++?

This is a hypothetical question. If object orientation is required and programmer is familiar with C++, it is always preferred to go with C++. It is by the very basic intuition, C++ is advanced than C++ and hence would be preferred by that way too.

**5.** What are the three basic concepts of Object oriented programming (OOP)?

The three basic concepts of OOP are:

i. Encapsulation
ii. Inheritance
iii. Polymorphism

**6.** What is a class? What is the relationship between class and object?

A class is a group of related objects aligned together. Or, multiple objects together form a class. An object will be an instance of a class too. Members of the structure are public by default whereas members of the class remain private.

**7.** What is the difference between a class and a structure?

Class is always seen as an advancement of the structure. There are few major differences between them. Class uses the constructor and destructor. Structure does not use constructor or destructor.

**8.** What is the size of an empty class?

The size of an empty class is always 1 byte. An object can be created for a class even though it is empty. So, to hold that object, minimum amount of memory required is 1 byte. One can try running the below piece of code to get a clear visibility.

```
#include <iostream>
 using namespace std;
 class function_echo
 {
 };
int main(void)
{
  function_echo foo;
  cout <<"\n" <<sizeof(foo);
  return 0;
}
```

**9.** State a few object oriented programming languages.

1. Small Talk
2. C++ (not fully object oriented)
3. Java (pure object oriented language)
4. Python
5. Lingo
6. Perl (from Version 5.0)
7. Python

**10.** How is C different from C++? State 5 points.

| C | C++ |
|---|---|
| C is a procedural language, i.e., it is a POP. | C++ is an object oriented programming language. |
| Data is not so secured in the case of C programming whereas it is highly secured in C++. | Data hiding is supported C++ |
| C language it is fully function driven. | C++ is fully object driven. |
| C programming follows a top-down approach. | C++ follows a bottom-up approach. |
| Function overloading is not permitted in C. | Function overloading is supported in C++ |

**11.** How is C++ different from Java? State 3 points.

| C++ | Java |
|---|---|
| C++ support operator overloading. | Java does not support operator overloading. |
| C++ has the support for pointers. | Java does not have the concept of pointer. |
| Garbage collection is not supported in C++ | Java has got a support for garbage collection, which will automatically free the memory. |

**12.** What are basic features supported in object oriented languages?

Polymorphism, virtual functions, inheritance, inline functions, etc., are supported in object oriented languages.

**13.** What is an object?

Anything that can be seen in real world can be referred to as object. Object can also be referred as instance of class. Alternatively, a group of similar objects forms a class.

14. How can someone identify something as an object?
    There are three basic things to be remembered about an object. It has
    - Attributes
    - Behaviour
    - Identity.

    With these three one can identify an object.
15. What is encapsulation?
    Enfolding of state and behavior into a single unit, this process is known as encapsulation. It can be achieved using a keyword called Class.
16. What was the older name of 'C++'?
    C++ was originally called C with classes. Later on it got changed as C++.
17. Name some real time objects and give their attributes.
    A table for an instance. It has a name called Table. It will have a color. It will have size and behavior. Any object will have these, by default.
18. Is there any difference between 'printf' and 'cout'?
    Printf () is a function whereas cout is not a function. Cout depends on the operator overloading concept by using << operator whereas printf is no way like that. Cout is always regarded as more extensible in nature as it is easier to work with any data type. Also, the formatting and related stuff remain easy with respect to cout where it is complex in printf.
19. What is IOSTREAM? How is it used?
    IOSTREAM is like stdio.h in C programming. IOSTREAM defines the standard input/output services for the C++ program. One cannot use cout or cin without having the iostream header file included.
20. Can a structure be a member of a class?
    Yes.
21. Can a structure be defined inside a class? If so what will be its scope?
    Yes the structure can be the member of the class and it should be in the public scope so that the return type is publicly visible. Otherwise will throw an error.
22. Can a file object be a member of a class?
    No.
23. Can a C++ program have two main functions? If so which will be executed first?
    No. The compiler won't compile if it has 2 main methods. It throws error saying "redefinition of 'int main () ' ".
24. What are predefined functions? Where is it defined?
    The predefined functions are some standalone functions of C++ Library (like cmath or cstdlib) that are declared in the header files using pre-processor directives. Predefined functions are listed below:

- sqrt
- pow
- abs
- exit
- assert

**25.** Can a program be executed without header files included?

No, it throws an error. The program should include header files which are like standards to define for writing a program. Therefore, the program should always begin with the header files.

26. How to compile different files each with a class, and an independent file with 'main' class?

In C++ we save classes or main function in separate files, compile them separately and finally link them to get executable as follows:

- g++ -c grace.cpp
- g++ -c main.cpp
- g++ main –o grace.o – o samples

  and execute the samples files as follow
- /samples

**27.** What modifiers are available in C++?

The modifiers in C++ are

- Signed
- Unsigned
- Long
- Short

**28.** Can a variable declaration have a modifier and variable name alone skipping the data type?

Yes, the variable declaration can have modifier and can have data type skipped for that variable. The variable takes integer data type by default.

**29.** Can two classes with same name exists in different scope?

No two classes with the same name can exist with same name even in different scope. It throws an error "redefinition of 'class name' error during compilation.

**30.** Can two classes with same name and same object name (from different scopes) interact with each other?

It throws an error saying "redefinition of class 'class name'" for having same class name even though with different scope calling from same object.

**31.** Can a class have a function named main()?

A class cannot have a function name "main". The compiler would throw error for this.

32. Is class a datatype? If so, why it is preferred?
Yes, class is user defined data type. Class ensures good reusability. Also, the name of the class can be decided by the programmer (though few restrictions like main cannot be the class name).

33. Can an access specifier be defined twice (i.e., can a class have two 'public' or 'private' or 'protected' part)?
Yes, a class can have access specifier defined twice. The program execution takes place correctly without any error. But, this does not serve any purpose and can be avoided.

34. How many header files are present in C++ standard and how many of them are borrowed from C standard?
33 header files are specific to C++ and 18 files are borrowed from C standard.

35. Is it possible to create a header file by a user? If so, how?
Yes, it is possible to create a header file. One can include the created header file by using # include "yourfile.h". Standard header files as iostream will be included as # include <iostream>.

36. Can a 'this' pointer of a class be used in another unrelated class?
The "this pointer" of one class cannot be used by other class. The other class is totally unaware of this keyword called 'this' and hence is restricted.

37. What is a pointer to an object? How can the members be accessed using this pointer to the object?
Pointer is a variable that holds address of any given object or data types. The member of class is accessed by "this" pointer as follows:

```
this->pfun()
(*this).pun()
```

->is also called array of pointer

38. What is the restriction imposed on the programmer while using the "this" pointer?
"this" pointer cannot be used with static members of class. It will flash a linker error!!

39. Who all can access the protected members of a class?
Protected member can be assessed by their own class and any derived class but cannot be accessed outside the derived class.

40. What is a 'this pointer'? How to use it?
'this pointer' is a pointer that points the current object to its own address space. It is used by using a keyword called 'this' . Suppose that there are 5 objects (ob1,ob2,ob3,ob4,ob5) for a class named "Area_Volume". The class has methods for calculating volume of cone and area of circle. The five objects keep accessing the member function one after the other for various purposes. the "Area_Volume" class wants to keep track of the objects that currently access the member functions. This is possible with the help of a pointer called "this" in c++. It holds the address of the object that called the member functions. Whenever a method call occurs, the address of the object is passed to 'this' pointer automatically.

**41.** Can a constructor of one class be present in a different class? Will there be some errors in this?

The constructor of base class is getting automatically available in the derived class due to the very property of inheritance and it would not throw any error.

**42.** Can a subclass contain the constructor of the outer class?

The subclass contains the constructor of the outer class and it is very obvious.

**43.** How many overloaded constructors can a class have?

There is no such limit. Overloading can happen any number of times. One should make sure that the parameters are different in terms of number or type or order.

**44.** Can an overloaded function be called explicitly from another member function?

Overloaded function is nothing but a member function. Therefore a member function can call other member function.

**45.** How many parameters can a constructor have? Can it be of different types?

A constructor can have any number of parameters. It can also be different in type and order of parameters.

**46.** Even when there is an explicit simple constructor defined, will the implicit constructor be called and then the explicitly defined constructor is called?

Implicit constructor gets called only when there is no explicit constructor defined.

**47.** Does the implicit constructor initiate all object variables value to 0 or null?

Depending on the data type the implicit constructor would initiate the object variables to 0 or NULL.

**48.** What is the primary work of a constructor and destructor?

A constructor is called by compiler when an object is instantiated. Destructor is called when the object gets deleted or the object gets out of scope. Both the constructor and destructor behave like a normal member function of the class. As the name suggests, the compiler allocates memory for the object while calling constructor, whereas the memory is refreshed or cleaned while destructor is called.

**49.** Can a destructor be recursed?

No. It is not possible.

**50.** Can a destructor be called explicitly by an object?

Yes it can be called. But it has no significance.

**51.** What is the amount of memory required to have int pointer and float pointer?

Any pointer in C and C++ will need only 4 bytes of the memory. Be it integer or float or character, it does not matter.

**52.** Why destructors are not used in Java while they found lot of usage in C++?

Since garbage collection feature is enabled in Java, there is no need for someone to use destructors.

**53.** C++ is always spotted as object driven while C is seen as function driven. Explain this.

C++ is fully object driven, i.e., objects are the fundamental building blocks. But, in the case of C language, it is fully function driven. Functions act as the fundamental blocks in C.

54. What happens in an object declaration?
    When an object is declared, the constructor in the class will be called that will be performing the initialization of data members.
55. Can an object be passed as an argument to functions? (Yes/No).
    Yes, very well. (Few examples are covered in the sections earlier)
56. How to call the static members of the class? Can objects be used for doing so?
    Static members can be called using the class name itself. It is done with scope resolution operator. For example, let fun() be a static member of the class c1. Then call the function like c1::fun(). Never try using the object to call static functions. It will throw compile time error.
57. Explain the mechanism of passing and returning objects to and from the functions.
    When an object is passed and returned, a copy of the object is passed to and from the function. Call-by-value is the mechanism followed in this scenario.
58. What is inheritance?
    Inheritance is the ability of one class to inherit properties from another class.
    For example: A derived class inherits the properties of a base class.

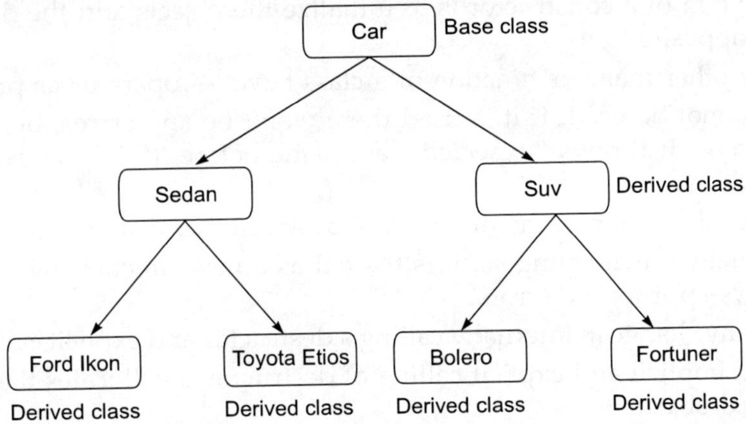


Here, CAR is base class or super class or parent class. From that base the inheritance will happen and will create the derived classes SEDAN and SUV.

59. What are all the types of classes?
    There are five types of classes. They are:
    - Base class
    - Derived class
    - Nested class
    - Local class
    - Incomplete class

**60.** What is a local class?

Local class is a class type declared within a function definition. A local class type name can only be used in the enclosing function. All the methods of the local class have to be declared only within the class.

**61.** How is scope resolution operator useful?

Scope will allow a program to refer an identifier which is declared in the global space with having the same identifier hidden in the local scope.

**62.** How will you compile a C++ program in gcc?

g++ -o <executable_file_name><cpp_file_name.cpp>

**63.** How is Inline function useful?

In case a function is getting called so many times in a particular program it has to waste a lot of time in passing the control, exchanging values, etc., Instead of wasting time in this way, if an inline function is called, the compiler inserts the full code (function) at the calling place and the time will be saved.

**64.** What is the difference between inline function and macro?

Inline functions are seen as a request which may or may not be respected by the compiler and all the inline requests are handled by the compiler. But, macro is handled by preprocessor and it gets expanded even before the compilation.

**65.** What is the role of a constructor in a C++ program?

The sole aim of a constructor is to initialize the objects and the destructor stands exactly opposite to it.

**66.** Can any other member function of a class have '~' operator as prefix?

No. It cannot be used. If it is used there won't be any erros, but when calling it errors come. It throws "expected class-name before '(' ". This treats the call as a call for destructor.

**67.** Can a subclass contain the outer class' destructor or constructor?

No. Actually when trying so, it is treated as a new function without return type. So throws appropriate errors.

**68.** Is there any change in internally calling a destructor and explicit call of a destructor?

No. Both implicit and explicit calling of destructor de-allocates the memory of the associated object.

**69.** Can a variable have multiple storage classes?

No. It gives error "conflicting specifiers in declaration".

**70.** What is the significance of using 'auto' storage class?

There is no significance for the 'auto'. Actually all local variables are by default 'auto' so it has no significance in usage. The significance of being 'auto' is to release the unwanted memory held after usage.

**71.** What is 'register' used for?

Using 'register' in declaration requests the compiler to store the variable in CPU register instead of RAM. So it stays in memory even after completion of the program and is faster. But allocating register completely depends on the processor.

**72.** How is 'static' significant?

It helps in creating a variable that acts as a common attribute for all objects of the class. Additionally, it helps in having a 'count' of objects and constant valued attributes.

**73.** How will encapsulation of a class take place when it has 'static' members?

When an object it created, it gets initial 1 byte memory. Then the compiler allocates memory for all non-static members of the class and links it to the memory of the object. Also the remaining 'static' member's memory is also associated to the particular object.

**74.** Can a static function use a non-static global variable?

Yes. It can use any variable outside class scope. But inside class scope should use only static variables.

**75.** Can a 'private' static member be called outside? Will the same syntax work?

Consider this code:

```
#include<iostream>
using namespace std;
class base
{
static int b;
}S;
int base::b = 8;
int main()
{ cout<<base::b;}
```

It throws error 'int base::b' is private. Private static members cannot be accessed.

**76.** How is storage class different from modifiers?

Storage classes are used to modify the storage characteristics of the variable. While modifiers are used only to modify the usage characteristics of the variable. All variables with or without modifiers have the same storage property. All are 'auto' by default.

**77.** Can a variable have a storage class and modifier together?

Yes. It can have both storage class and modifier. As they change independent properties, it is allowed.

**78.** Can a variable declaration have a storage class and variable name alone skipping the data type?

No. In declaration after storage class compiler expects a type name. On such usage variable name will be regarded as type name and appropriate errors will be thrown.

**79.** Can a storage class be specified after the datatype in declaration?

Yes. The compiler automatically maps the keywords and competes the declaration. It doesn't have any change in the default way of declaration.

**80.** What are the access specifiers supported in C++?

There are three access specifiers supported in C++. They are:

- Public – The members that are declared as public are fully accessible from outside that class through an object of that class.
- Protected - The members that are declared as protected are accessible from outside the class, but only in a class derived from it.
- Private - These members are only accessible from within the class. No outside access is allowed.

**81.** Differentiate between call by value and call by reference.

In call by value, the original values will remain the same even after the function is called because the copy of the variables is being passed. But in the case of call by reference, the original values will be altered after the function is called because address is being passed.

**82.** How can references be used in C++?

References can be used to refer one address with multiple names. For an example, one person can be called with multiple names. This feature is referred to be as aliasing.

**83.** What is a friend function? When will it be needed?

When a non-member function is declared as a friend it would be easier for the function to access the private members. Likewise, when a class itself is declared as a friend, the friend class can then have access to the private data members of the class which made a friend.

**84.** How will you allocate and deallocate memory in C++?

Allocation of memory can be carried out with new operator and de-allocation can be carried out by delete operator.

**85.** What is the return value of a new operator?

When the new is used to allocate memory, if it has been allotted properly, address will be returned.

**86.** What is inheritance?

Inheritance is an object oriented feature. It is gaining features from one class to other class. The features gained are the data members and member functions of the base class. It is the act of giving same properties to lower class.

**87.** Give some real time examples of inheritance.

- Son gaining business tactics from father.
- Students gaining tricks and knowledge from teacher/master.

**88.** What are the modes of inheritance?

The modes of inheritance are:

- Public
- Private
- Protected

89. What is the syntax for inheritance?
class derived_class : mode base_class1 , mode base_class2, .....

90. In inheritance, when a derived class has a normal member and another inherited member also has the same name, if the derived class object calls the member then which will be called? The base variable or derived variable?
Consider this program:

```
#include<iostream>
using namespace std;
class base
{ public:
  int b; }S;
class dev : public base
{ public:
  int b; }D;
int main()
{ D.b = 5;
cout<<D.b; }
```

Here D.b will access the original variable and not the inherited variable. To access the inherited variable, the base class name must be used.

91. What is the syntax for 'this' operator?
((class_name *)this)->member

92. In 'public' mode of inheritance, where will the 'private' members of the base class present in the derived class?
'private' members of the base class will not be derived into derived class in any mode of inheritance.

93. How to inherit the 'private' members of a class?
'private' members of a class can never be inherited.

94. How to inherit 'static' members of a class?
'static' members are inherited as any normal variable. But for normal variables a copy of it will be created in the derived class, but here access to the original variables will be given. It should be accessed with the appropriate class name.

95. Will a 'static' function be inherited as a 'static' function?
Yes. A 'static' member will always be 'static'.

96. Can an inherited 'static' function be called with 'derived class ::' this syntax?
Yes. For 'static' functions it can directly be called from derived class unlike 'static' data members.

97. What happens when the inheritance access specifier is not specified in inheritance, i.e, class derived_class : base_class ?
The class will be inherited in 'private' mode. The mode of inheritance is 'private' by default.

**98.** Can a class inherit an undefined class that is not even declared?

No. Using unknown classes throws error "invalid use of incomplete type 'class class_name' ".

**99.** Can a class inherit an undefined class that is declared alone?

No. Even though it is declared still it is incomplete. It throws the same error.

**100.** Can inheritance be done with predefined or other user defined data types?

No. Inheritance is an object oriented feature. It is allowed only among classes.

**101.** Explain the role of '::' operator with respect to inheritance.

When a local member and inherited member have the same name collision occurs. To call the inherited member the class name along with '::' should be used. In general it is used for hierarchical access and variables in higher scopes.

**102.** Explain single inheritance. Give a real time example.

It is a single derived class inheriting from a single base class, e.g., son learning business tactics from father.

**103.** Explain multiple inheritance. Give a real time example.

It is a single derived class inheriting from more than one base class simultaneously, e.g., students gaining knowledge from teachers.

**104.** Explain multilevel inheritance. Give a real time example.

It is a derived class inheriting from one base class, which in turn has inherited from a super base class. Thus, inheriting in a chain, e.g., passing business across generations from grandfather to father and from father to son.

**105.** Explain hierarchical inheritance. Give a real time example.

It is multiple derived class inheriting from one common base class, e.g., family secrets passing from father to two sons.

**106.** Explain hybrid inheritance. Give a real time example.

Any combination of above inheritance types leads to hybrid inheritance, e.g., school management assigning work from top to bottom. It has multiple teachers under principle, but one common assistant.

**107.** What is virtual inheritance?

Hybrid inheritance is also called virtual inheritance. It is called so because of the usage of 'virtual' keyword in it. When there is a chance of duplication of a super base class attributes in any derives class 'virtual' should be used. It makes only one copy attributes and skips other copies based on the hierarchy of the attributes.

**108.** Is 'virtual' keyword mandatory in virtual inheritance?

No. When the conflicting base class is empty there is no chance of duplication. So it can be skipped. 'virtual' is not mandatory in the syntax. But it has no significance. So indirectly 'virtual' keyword becomes mandatory.

**109.** What is the type of inheritance if it follows a pattern that is combination of available inheritance methods.

Any abnormal inheritance pattern falls under the roof of 'hybrid' inheritance.

**110.** Is there any physical limit on the maximum number of classes that a class can inherit? (how may base classes can a derived class have?)

No. A class can inherit any number of classes provided the memory (RAM) is available.

**111.** What is polymorphism?

Adding functionalities to the existing attributes is called polymorphism. Polymorphism means 'different forms'. It involves operator overloading and functional overloading.

**112.** What is operator overloading?

Adding new functionalities to the existing operator is called operator overloading. But the original functionality remains unchanged.

**113.** Is it possible to remove the default functionality of an operator using operator overloading?

No. There is no way to remove the original functionality of the operator.

**114.** Can an overloaded operator be used by its inherited classes?

Yes. If the operator definition is in 'public' scope, it will be inherited by its derived classes also.

**115.** What is function overloading?

Giving multiple definitions to a class member function is called function overloading.

**116.** What are the rules for overloading?

- The name of the function should not be changed.
- The return may or may not be varied.
- The number of arguments may or may not be varied.
- The type of the arguments may or may not be varied.
- Either one or multiple from the following must be varied: return type, type of arguments, number of arguments.
- Changing the argument name doesn't have an effect.

**117.** Does redefining a function by just changing the name of the arguments belong to function overloading?

No. renaming the arguments doesn't change the internal signature of the function.

**118.** Does redefining a function by just changing the order of types of the arguments belong to function overloading?

Yes. Changing the order of data types of the arguments creates a new signature of the class. Hence, it is an overloaded function.

**119.** Is it necessary to vary the number of arguments for an overloaded function?

No. If any either return type of the function or type of arguments are varied, then it is not necessary to change the number of arguments.

**120.** Where do we use operator overloading? How is it different from function overloading?

There are chances where a single operator may be used for two different situations as follows:

- ✓ 5 + 3 = 8
- ✓ Shri + ram = Shriram

Here + is the operator used to add two numbers and in the second case, it is used as a concatenation operator. This is the purpose of the operator overloading and function overloading will let two different functions to have same function name.

**121.** Can sizeof operator be overloaded? What are the other operators which cannot be overloaded. List them all.

Sizeof cannot be overloaded. There are few other operators which cannot be overloaded and they are listed as follows:

- ✓ . (Member Access or Dot operator)
- ✓ ?: (Ternary or Conditional Operator)
- ✓ :: (Scope Resolution Operator)
- ✓ .* (Pointer-to-member Operator)

**122.** Define single inheritance with appropriate pseudocode.

When one derived class is inherited from one base class, it is referred to be a single inheritance.

Pseudocode:

```
class base_class // This is base class.
{
   public:
             --------- // Code goes here.
};
class derived_class: public base_class
//Deriving class from the base and single inheritance is achieved here.
{
   public:
             --------- // Code goes here.
}
```

**123.** Explain multiple inheritance with an example.

When one derived class is derived (inherited) from multiple base classes it is referred to be a multiple inheritance.

```
class base_class_1 // Base class_1.One of the base classes.
{
   public:
             ----- // Code goes here.
};
class base_class_2 // Base class_2. Second base class.
```

```
{
  public:
        ----- // Code goes here.
};
class derived_class: public base_class_1, public base_class_2
// 2 base classes are now used derive one derived class and it is called multiple inheritance.
{
  public:
        ----- // Code goes here.
}
```

**124.** Explain multilevel inheritance with an example.

A class will be used to create the derived class. For example, A à B à C à D, where A is the base class, B being the first derived class, C being the second derived class, but derived from B and so on. This is called multi-level inheritance.

```
class base_class_1 // Base class_1. It is to be inherited.
{
  public:
  int x, y;
};
class derived_class_1: public base_class_1 // deriving from base class.
{
  public:
  int z;
}
class derived_class_2: public derived_class_1 // deriving from derived class_1.
{
  public:
  int z;
}
class derived_class_3: public derived_class_2 // deriving from base class_2.
{
  public:
  int z;
}
```

**125.** What is function overriding?

One can redefine a function declared in a base class to have the same function with a different usage in a derived class

**126.** What is the difference between function overriding and function overloading?

For function overriding:

1. The definition of the function being overridden must have same method name.
2. The base class function and the overriding function must have same data type.
3. The base class function and the overriding function must have same argument list.

Whereas in function overloading the signature (argument type and number of arguments) need not be the same. In fact, they are expected to be different at least in the type or number of arguments.

**127.** What is the use of templates in C++?

If a programmer is asked to write a function that can add two integers can the same function be used for adding one integer and one float? No. Eventually, the programmer will end up in writing 2 different functions that can handle the question raised. Templates will help the programmer to avoid such problem. It is very generic and a single function will be able to handle both the cases mentioned above with templates.

**128.** What is a virtual function?

Simple, it is used to give different meanings for one function. It might seem like calling one function, but the call could have been raised for a different function. Virtual keyword should be used for it.

**129.** Why cannot the constructors be declared virtual?

The virtual keyword will convey the compiler that the function call shall be made only during the runtime and not during the compile time. So, when coming to constructor the object creation itself is dependent on the constructor and cannot wait. So, it would be inappropriate to make a constructor virtual.

**130.** How is pure virtual function useful?

In a base class, if a function is declared a pure virtual function, the compiler will now make it mandated for the derived class to override them.

**131.** What is an abstract class?

In other words, a class which has pure virtual function can never be used to get objects out of them. The class is then referred to be an abstract class.

**132.** Why should someone go for virtual base class?

When a class is derived out of two derived classes, then there will be two copies of the base class components. Then there will be an ambiguity to decide which copy is to be used. So in this case virtual base class concepts are to be used.

**133.** What is early binding? One simple example where it could happen?

A function call is resolved at compile time is called early binding and any direct function call is a simple example for the early binding.

**134.** Write on the access rights with all the kinds of inheritances.

The following diagram is sufficient to get the clarity.

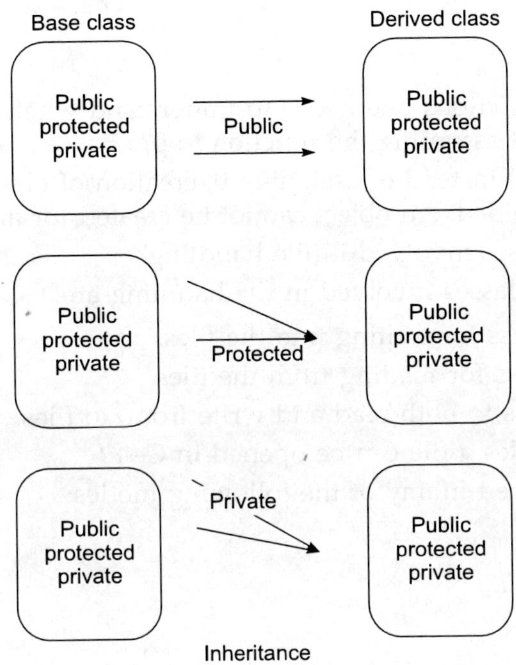


**135.** How can someone make a function pure virtual function?

(a) Option one, the function can be assigned to 0 and hence it becomes pure virtual function.

```
#include <iostream>
using namespace std;
class Base_class
{
public:
virtual void show() = 0;
};
```

(b) The below given example is not a mostly preferred way. But still, it would work like pure

```
virtual function.
#include <iostream>
using namespace std;
class Base_class
{
public:
virtual void show()
{ // Nothing inside. Hence, it is also performing pure virtual function's
functionality.
```

```
        }
    };
```

**136.** What are the restrictions one needs to understand when making a function pure with { } instead of assigning the function to 0?

When one uses { } instead of assigning 0, creation of object is permitted which is actually to be avoided. An object cannot be created for an abstract class.

**137.** What are the classes involved in file handling?

The three major classes involved in file handling are:

- ✓ ofstream: Class for writing into the files.
- ✓ ifstream: Class for reading from the files.
- ✓ fstream: Class to both read and write from/to files.

**138.** What are the modes a file can be opened in C++?

A file can be opened in any of the following modes:

- ✓ ios::app
- ✓ ios::ate
- ✓ ios::binary
- ✓ ios::in
- ✓ ios::out
- ✓ ios::nocreate
- ✓ ios::noreplace
- ✓ ios::trunk

**139.** What are the uses of the following functions? A. seekg (), B. seekp() C. tellg () D. tellp ()

- ✓ seekg () will move the get pointer (input) to a specified location.
- ✓ seekp () will move the put pointer (output) to a specified location.
- ✓ tellg () will fetch and give the current position of the get pointer.
- ✓ tellp () will fetch and give the current position of the put pointer.

**140.** Brief on the following functions related to file handling. A. eof () B. fail () C. bad ( ) D. good.

| Function | Meaning |
|---|---|
| eof() | Returns true with a non-zero value if end of file is encountered while reading; otherwise return false (zero) |
| fail() | Returns true when an input or output operation has failed |
| bad() | Returns true if an invalid operation is attempted or any unrecoverable error has occurred. |
| good() | Returns true if no error has occurred. |

**141.** What are the types of files one can list when asked?

✓ Text file

It could be a typed document which could have series of alphanumeric characters or symbols. One thing to remember is, every line of a text file is terminated with an EOL (End of Line) character.

✓ Stream

Sequence of bytes is normally referred to be a Stream.

✓ Binary files

A binary file will contain data encoded in binary form for computer storage and processing purposes. Only the processor can understand it and not a human. Because, it would be in 1s and 0s.

**142.** What is the use of Void pointer in C++?

Let's take a bottle; if you pour water into it, it is a water bottle. If you pour juice into it, it's a juice bottle. A pointer is also similar. If a pointer holds the address of a float or int or char, etc., it is referred to be a void pointer. When it holds the address of a float only, it becomes float pointer. When it holds the value of integer alone it becomes an integer pointer. It is very flexible. The most important point to remember is that casting has to be done to make the void pointer a valid data type pointer

```
#include<iostream>
using namespace std;
int main ()
{
        int a = 9;
        double b = 10.676;
        void *p; // void pointer

        p = 8a;
        // Here it is made as a integer pointer. Means p can hold integer address.
        cout <<"\n" << "Integer Pointer "<< *(int *)p;
        // Casting becomes inevitable. Making the void pointer as int pointer.

        p = 8b;
        cout <<"\n" << "Double Pointer "<< *(double *)p;
        // Casting becomes inevitable. Making the void pointer as double pointer.
        getchar ();
        return 0;
}
```

**143.** What is the use of null pointer in C++?

When an uninitialized pointer is used, there could be a possibility that it points to some garbage value. To avoid that it is better to initialize the pointer to 0. A pointer when initialized to 0 is referred to be a null pointer. It points to nowhere. A simple example as ever would do. Reader can refer to the code below for getting a clear understanding.

```
#include<iostream>
using namespace std;
int main ()
{
    int *p;
    p = NULL; // NULL is the key word here.
    cout<<p; // This pointer p is initialized to a NULL and it will be pointing to 0.
    getchar ();
    return 0;
}
```

Execution Result:

**144.** What is friend class? How is it useful?

When there is a private data member in a class, access is restricted for the other classes, i.e., no other class can access them. So the best way to get this done is to make the second class, which needs to access the private data members of the first class, a friend. Since both the classes are friends, it becomes easier. To enable the access easier, we make a class friend to another class.

**145.** What are the storage qualifiers supported in C++?

There are four standard qualifiers supported in C++ as in C programming. They are:

- auto
- static
- register
- extern

# 7 ELECTRICAL MACHINES

Machine is the most important area irrespective of any field of engineering. This helps us not only to face interview but few of its basics are helpful in many other areas of engineering and also in day-to-day life. This machines part covers areas like dc machines, ac machines, motors, generators, transformers, etc.

1. **What is Motor? (L1)**

   It is a device which converts electrical energy to mechanical energy.

   For example – FAN which when switched on converts electrical energy into rotational (mechanical energy).

2. **What is magnetic field? (L1)**

   The area around the magnet where its magnetic intensity is felt.

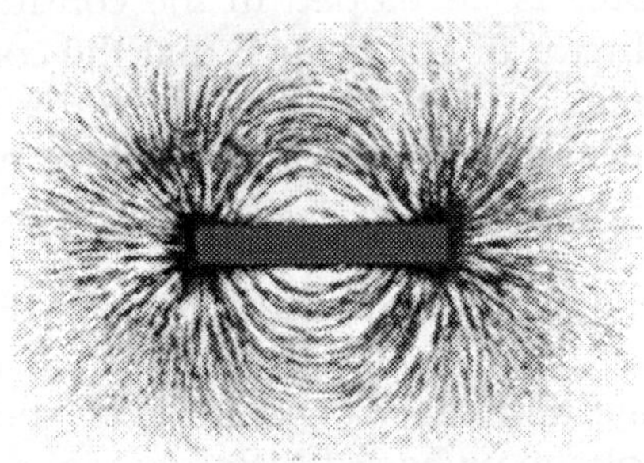

3. **What is EMF? (L1)**

   EMF stands for Electromotive Force. It is the potential difference (voltage) generated between two electrodes.

4. **What is electromagnet? (L1)**

   It is a material which appreciates magnetic property when it is energized.

   Example: Iron

5. **Define torque. (L1)**

   Torque is defined as a force that produces or tends to produce rotation.

   Torque of a motor is calculated by

$$T = K\phi a$$

where $\phi$ the flux/pole in weber, $K$ is a constant depending on coil geometry, and $I$ is the current flowing in the armature winding.

6. **What is Faraday's law of electromagnetism? (L1)**

   It gives the relation between electric circuit and magnetic circuit. It states that whenever a conductor is placed in varying magnetic field an EMF is induced across the conductor.

7. **What is residual magnetism? (L1)**

   Residual magnetism is a property of a material which is defined as the amount of magnetization remains even after removing the external magnetic field from the circuit. In other words, it is the value of the flux density retained by the magnetic material.

8. **What is coercive force? (L1)**

   The opposing magnetic intensity that must be applied to a magnetized material to remove the residual magnetism.

9. **How a motor works or state the working principle of a motor? (L1)**

   When a current carrying conductor or a coil is connected to a DC source which is placed in a magnetic field, there will be an interaction between electric and magnetic fields. Therefore, a force is developed in the conductor or coil. The force on the two sides of the conductor or coil is such that the conductor or coil starts to move in the direction of force. Thus, the electrical energy in the conductor is converted to rotational/mechanical energy (Faraday's law of electromagnetism).

10. **Define counter voltage? (L1)**

    Counter voltage is also called back EMF. Back EMF in a motor is defined as the voltage that is produced due to change in the flux. Thus, the voltage produced is opposite to the voltage that causes current flow through the conductor which is referred as counter-voltage or back EMF.

11. **What are the components of a motor? (L1)**

    1. Yoke
    2. Poles
    3. Armature winding
    4. Field winding
    5. Commutator
    6. Brushes

12. **What are types of DC motors? (L1)**

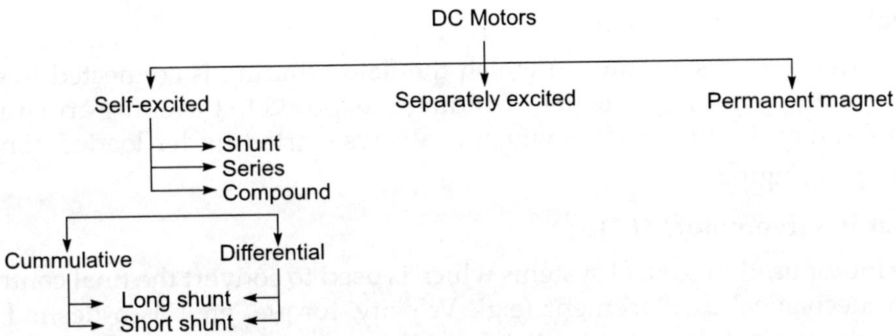


**13. What are types of AC motors? (L1)**

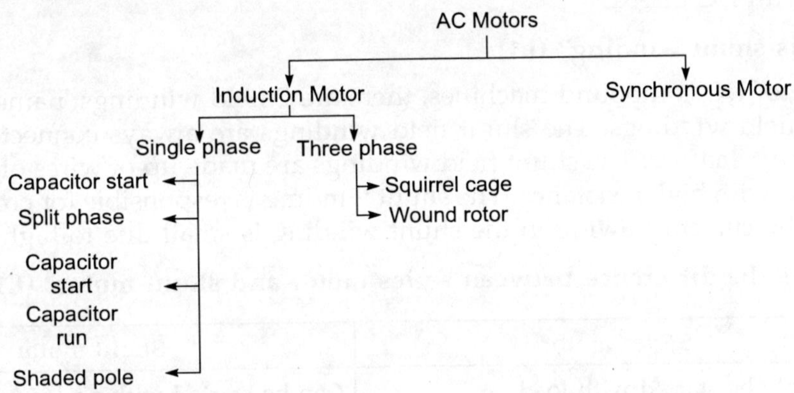


**14. What is active power? (L1)**

In simple terms it is defined as power working on the load.

**15. What is back pitch? (L1)**

Back pitch is the distance between top and bottom coil sides of a coil measured around the back of the armature, away from the commutator.

**16. What is a brush and commutator? (L1)**

**Commutator:** The armature coil in the motor is connected to the commutator which is a cylindrical body. Its main function as far as the DC motor is concerned is to commute or relay the supply current from the mains to the armature winding housed over a rotating structure through the brushes of DC motor.

**Brush:** Brushes used in a motor are made up of carbon or graphite which conducts current between stator (windings) and rotor (moving part).The brushes are used to relay the current from external circuit to the rotating commutator from where it flows into the armature winding.

**17. What is series motor? (L1)**

DC series motor is a motor in which the field winding is connected in series with the armature winding. The field winding is exposed to the entire armature current. It is a self-excited DC motor which is always started under loaded condition.

Example: Engines.

**18. What is servomotor? (L1)**

The motor used in control systems which is used to convert the final control element into mechanical displacement (e.g., Velocity, torque, etc.) as a desired output. It has low moment of inertia for the rotor, linear speed-torque curve having negative slope. These motors are built specially to handle frequent start and stop. It can be used with DC or AC.

**19. What is shunt winding? (L1)**

In case of DC compound machines, there are 2 field windings, namely, shunt and series field windings. The shunt field windings are always connected in parallel with the armature. The shunt field windings are made up of wires of smaller cross sections with high resistance. The shunt winding is responsible for creating of main flux. The current flowing in the shunt winding is small due to high resistance.

**20. What is the difference between series motor and shunt motor? (L1)**

| Series motor | Shunt motor |
|---|---|
| Can only be started with load | Can be started with no load |
| High starting torque | Starting torque is less than series motor |
| Less number of turns in winding | More number of turns in windings |
| Has the field windings connected in series with the armature. | Has the field windings connected in parallel with the armature. |
| Variable speed motor. | Constant speed motor |

**21. Why a series motor should start with load condition? (L2)**

In a series motor the series winding which has less number of turns, during the start with no load or fewer loads can generate a moderate flux. But inversely during starting, the current flow will be very high. Because of the high starting current in no load condition, the motor will rotate with extremely dangerous speed causing the damage to motor. So in order to limit the starting current, a load is applied to prevent the damage to motor.

**22. What is equalizer? (L2)**

In case of a DC motor, circulating current is developed since unequal values of EMF are generated in parallel paths of the armature winding. This circulating current doesn't perform any useful work and also generates excessive heat. In order to overcome the circulating current, equalizers are introduced in lap wound armatures. They are low resistance copper conductors. These equalizers normalize the potential difference inside the winding and also the current flowing through these equalizers will bypass the current from flowing through the brushes.

**23. What is a flywheel? (L2)**

Flywheel is a disc that is used to store rotational energy. It is a rotational mechanical device. In general they are DC compound motors in which the amount of energy stored is proportional to the square of its rotational speed.

**24. What is Hopkinson's test? (L2)**

Two identical mechanical machines are coupled together in which one machine acts as a motor and the other as a generator. In this connection, the generator will provide required input power to the motor. The supply voltage is to compensate the losses in these machines. This test is conducted to determine the efficiency of both the machines.

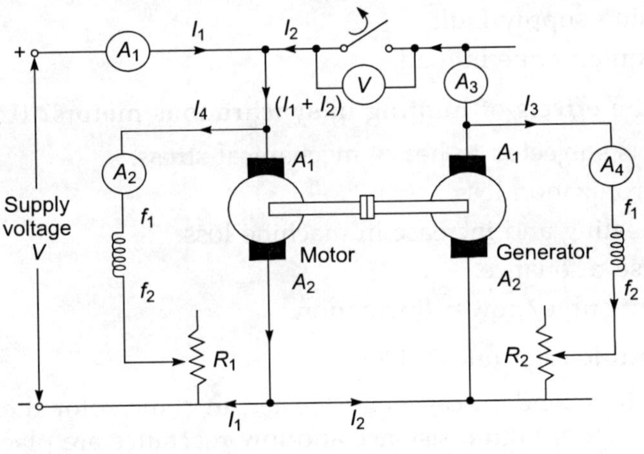


**25. What will happen when a DC motor is connected to AC source? (L2)**

The motor will start and rotate in reverse direction, because of this excessive heat is generated which burns the coil.

**26. What is capacitor start induction motor? (L2)**

This is a single phase induction motor in which a capacitor is connected in series along with the starter winding. Since the one-phase induction motors are not self-starting, these motors are connected with a capacitor to create phase difference. We are familiar with this fact that the current flowing through the capacitor leads the voltage. The winding is inductive in nature which the current lags behind the applied voltage. Hence, there is a phase angle difference between two currents produces resultant current which produces a magnetic field.

For example, conveyors, compressors.

**27. What is cogging? (L2)**

Cogging is a condition of an induction motor which occurs when there are equal number of slots in rotor and stator. If the slots are equal, then the motor will not start due to the coincidence of slot frequencies with harmonic frequencies which cause torque variations.

**28. What is hunting? (L2)**

In synchronous motors, oscillations are developed when it is suddenly loaded or unloaded. The speed of the rotor varies temporarily until the torque angle adjusts to new output requirements. In order to adjust the correct torque angle, the speed becomes constant. Hence, hunting is a condition in which the rotor revolves in forward and reverse direction at the average constant speed.

**29. What are the causes of hunting in synchronous motor? (L2)**

A. Load changes suddenly.

B. Field current changes suddenly.

C. Incoming supply fault.

D. Harmonic torque in load.

**30. What are the effects of hunting in synchronous motors? (L2)**

A. Rotor is subjected to heavy mechanical stress.

B. Synchronization loss.

C. Overheating and increase in machine loss.

D. Increase resonance.

E. Current surge/power fluctuation.

**31. What is double cage rotor? (L2)**

When two independent cages are located in same rotor then it is called double cage rotor. Bars of high resistance and low reactance are placed in the outer cage, and bars of low resistance and high reactance are placed in the inner cage. The outer cage has high 'reactance to resistance ratio' whereas, the inner cage has low 'reactance to resistance ratio'. A very low reactance is achieved by this construction. Thus, the double cage induction motor provides excellent speed torque characteristics during start-up.

**32. What is air-gap? (L2)**

There are two types of winding in transformers called primary winding and secondary winding. This primary winding is on the source side and secondary winding is on the output side. The gap between the two windings is known as air-gap.

**33. What is autotransformer? (L2)**

The autotransformers are the machines consist of only one winding which acts both as primary and secondary windings. The voltage difference between the two points acts as a potential difference which helps to get output at a certain point in winding.

**34. When autotransformer is more economical? (L2)**

They are economical when the voltage ratio between primary and secondary winding is smaller.

**35. What is Lenz law? (L1)**

It states that the direction of an induced EMF in a coil is in such way that it opposes the change producing it.

$$e = -Nd\phi/dt$$

**36. What is mmf method? (L1)**

This method of determining the regulation of an alternator is also called ampere-turn method or Rother's MMF method. The method is based on the results of open circuit test and short circuit test on an alternator.

**37. What are open circuit characteristics? (L1)**

- The output voltage is equal to the input voltage.
- Maximum voltage is measured in an open circuit.
- Zero current is measured in an open circuit.

**38. What are phasor diagram? (L1)**

The relationship between various quantities like applied voltage, core flux, induced EMF and current in the windings and voltage drops in the winding are expressed in a graphical notation called phasor diagrams. It is used to analyze the performance characteristics of electrical machines like generators, transformers.

**39. What is reactive power? (L1)**

Reactive power in AC circuits is the reactance in which the product $E_{rms} * I_{rms}$ is greater than the true power. The excess is called reactive power, and represents energy alternately stored and released by inductors and/or capacitors in the AC circuits.

**40. What is Scott connection? (L2)**

Scott connections are most widely used in transformers in which the three phase connection is converted to 2-phase connections. It needs a main transformer (center tapped), teaser transformer.

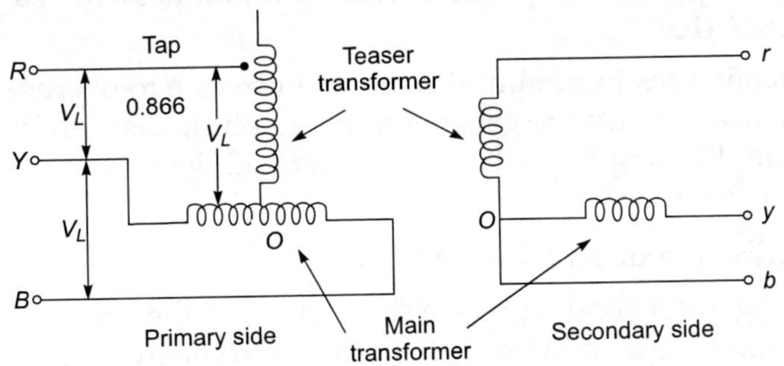


**Scott connection**

**41. What is generator? (L1)**

It is an electromechanical device which converts mechanical energy into electrical energy. It obeys Faraday's law of electromagnetism in which a conductor or coil placed in magnetic field experiences EMF. They are also called alternators.

**42. What are the types of generators? (L1)**

It is mainly classified into two types

- DC generators
- AC generators

Ac generators is again classified as:

- asynchronous generators
- synchronous generators

**43. Define the term voltage regulation of an alternator. (L2)**

The voltage regulation of an alternator is defined as the change in terminal voltage from no-load to load condition expressed as a fraction or percentage of terminal voltage at load condition; the speed and excitation conditions remaining same.

**44. How a synchronous motor can be used as a synchronous condenser? (L3)**

Synchronous motor is operated on overexcitation so as to draw leading reactive current and power from the supply lines. This compensates the lagging current and power requirement of the load making the system power factor to become unity. The motor does the job of capacitors and hence called synchronous condenser.

**45. What happens when the field current of a synchronous motor is increased beyond the normal value at constant input? (L3)**

Increase in EMF causes the motor to have reactive current in the leading direction. The additional leading reactive current causes the magnitude offline current, accompanied by the decrease in power factor.

**46. What is the normal value of slip of an IM operating at full load? (L2)**

It is 3 – 5.

**47. What would happen if a 3-phase induction motor is switched on with one phase disconnected? (L3)**

The motor continues to run but is likely to burn due to overheating since negative phase sequence currents are generated due to phase disconnection. This will cause motor failure. Heating and vibrations generated depend on the load of the motor at which it is running.

**48. State the principle of 3-phase IM? (L3)**

While starting, rotor conductors are stationary and they cut the revolving magnetic field and so an EMF is induced in them by electromagnetic induction. This induced

EMF produces current if the circuit is closed. This current opposes the cause by Lenz's law and hence the rotor starts revolving in the same direction as that of the magnetic field.

**49. What is meant by infinite bus-bars? (L2)**

The source or supply lines with non-variable voltage and frequency are called infinite bus-bars. The source lines are said to have zero source impedance and infinite rotational inertia.

**50. What is an ideal transformer? (L1)**

Transformer which has 100% efficiency is called an ideal transformer.

**51. What are the types of braking? (L1)**

- Regenerative braking.
- Plugging type braking.
- Dynamic braking.

**52. What is regenerative braking? (L1)**

Regenerative braking takes place whenever the speed of the motor exceeds the synchronous speed. This braking method is called regenerative braking because here the motor works as generator and supply itself is given power from the load, i.e., motors. The main criteria for regenerative braking is that the rotor has to rotate at a speed higher than the synchronous speed, only then the motor will act as a generator and the direction of current flow through the circuit and direction of the torque reverses and braking takes place. The only disadvantage of this type of braking is that the motor has to run at super synchronous speed which may damage the motor mechanically and electrically, but regenerative braking can be done at sub-synchronous speed if the variable frequency source is available.

**53. What is plugging type braking? (L1)**

In this method the terminals of supply are reversed, as a result the generator torque also reverses which resists the normal rotation of the motor and as a result the speed decreases. During plugging external resistance is also introduced into the circuit to limit the flowing current. The main disadvantage of this method is that here power is wasted.

**54. What is dynamic braking? (L1)**

In this method of braking the motor which is at a running condition is disconnected from the source and connected across a resistance. When the motor is disconnected from the source, the rotor keeps rotating due to inertia and it works as a self-excited generator. When the motor works as a generator the flow of the current and torque reverses. During braking to maintain the steady torque sectional resistances are cut out one by one.

**55. What is synchronous speed? (L1)**

The speed at which an induction motor will operate depends on the input power frequency and the number of electrical magnetic poles in the motor.

$$n = f(2/p)\,60$$

where

$n$ = shaft rotation speed (rev/min, rpm)
$f$ = frequency (Hz, cycles/sec, 1/s)
$p$ = number of poles

**56. Define slip? (L1)**

Slip of an induction motor can be defined as the difference between its synchronous speed and its speed at full load. The speed performance of an electric motor can be calculated by a method known as percent slip. The slip is essential for producing torque that increases with increasing load.

**57. Define crawling. (L1)**

It has been observed that squirrel cage type induction motor has a tendency to run at very low speed compared to its synchronous speed, this phenomenon is known as crawling.

**58. Define power factor. (L1)**

It is the ratio between the useful (true) power (kW) to the total (apparent) power (kVA) consumed by an item of AC electrical equipment

**59. What is power factor correction? (L1)**

Power factor correction is technique used to restore power factor close to unity, i.e., "1", in order to achieve economical power consumption, the power factor must be more or less equal to 1. But most of our industrial, household appliances are made up of inductance in which current lags voltage. Hence, the power factor is less than 1. In order to make it near to 1, we introduce capacitor banks. This is called power factor correction.

**60. What are the components of AC power flow? (L2)**

- Real power or active power (P), expressed in watts (W)
- Apparent power (S), usually expressed in volt-amperes (VA)
- Reactive power (Q), usually expressed in reactive volt-amperes (var)

**61. What are the speed control techniques of three-phase induction motor from stator side? (L2)**

1. V/f control or frequency control.
2. Changing the number of stator poles.
3. Controlling supply voltage.
4. Adding rheostat in the stator circuit.

**62. What are the speed control techniques of three phase induction motor from rotor side? (L2)**

1. Adding external resistance on rotor side.
2. Cascade control method.
3. Injecting slip frequency EMF into rotor side.

**63. Explain the solid state speed control? (L2)**

In order to control the speed of the motor to a precise value we use solid state speed control methods in which semiconductor components such as triac, and diac are used to control the flow of the current. Hence, the speed is controlled to accurate RPM levels.

**64. What are the types of winding? (L1)**

- simplex lap winding
- simplex wave winding

**65. What are the differences between lap winding and wave winding? (L1)**

In the wave winding, two ends of a coil are connected to a commutator segments that are approximately 360 degrees apart. This way all the coils carrying current in the same direction are connected in series. Therefore, there are only two parallel paths between the brushes, $a = 2$ independent of the number of poles. This type of winding is used in low-current, high voltage application

In one type known as simplex lap winding, the end of one coil is connected to the beginning of the next coil with two ends of each coil coming out at adjacent commutator segment. For a progressive lap winding the commutator pitch $y = 1$. A typical coil of $N$ turns for a simplex lap winding is shown in figure below:

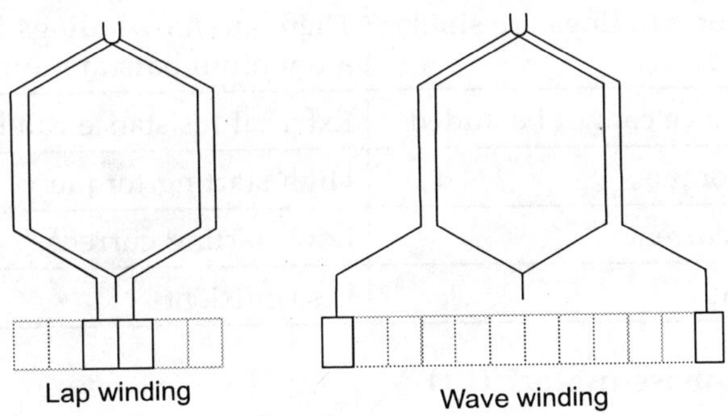


**66. What is skewing? (L2)**

Skewing is a technique to prevent cogging which means locking of motors.

**67. What are the advantages of DC motors? (L1)**

High starting torque

Excellent speed control capability

**68. Give some practical examples of high starting torque. (L1)**

Electric traction

Cranes

Elevators

Air compressor

Vacuum cleaner

Hair drier

Sewing machine

**69. What are reciprocating machines? (L1)**

Those are the machines which convert pressure, heat, pneumatic and hydraulic motion into rotating motion.

**70. Why DC shunt motor is a constant speed motor? (L2)**

In a DC shunt motor, the field winding and the armature winding are connected in parallel, hence both the windings will be connected to same voltage source. The input voltage source to the motor is constant, hence it generates constant flux. Speed of a motor depends on the flux generated. Thus, these motors are constant speed motors.

The current equation 'Ish' is given by Ish = $E$/Rsh, here $E$ = Input voltage which is constant, Ish = shunt current is also constant. Thus, the Ish is proportional to flux. Hence, the motor speed is almost constant.

**71. Comparison between squirrel cage motor and slip ring motor? (L1)**

| Squirrel cage | Slip ring |
|---|---|
| Rotor and stator windings are similar in constructions. | There are no windings in rotor, instead aluminium bars are terminated with slip rings. |
| External resistance cannot be added | External resistance can be added. |
| Low starting torque | High starting torque |
| High starting current | Low starting current |
| Highly efficient | Less efficient |

**72. What is split phase motor? (L1)**

It is a single phase induction motor that is called capacitor start motor. In these types of motors, a capacitor is connected in series with the startup winding. Thus, the circuit forms a LC combination which generates greater phase shifts. The starting winding is highly resistive; the running winding is inductive in nature. Thus, the resultant difference between these 2 windings generates a rotating magnetic field which rotates in one direction.

For example: SPIMs have low starting current and moderate starting torque.

- Fans
- Grinders
- Washing machines.

These motors are available in the size ranging from 1/20 to 1/2 kW.

**73. Comparison between star and delta connections? (L1)**

| Star | Delta |
|---|---|
| One end of each coil is connected to a common point. | Opposite ends of each winding are connected together to form a triangle. |
| Neutral point is taken from the common point, hence it is a 3-phase, 4-wire system. | No neutral point, hence it is called 3-phase, 3-wire system. |
| Line current is equal to phase current $I_L = I_{PH}$ | Line voltage is equal to phase voltage $V_L = V_{PH}$ |
| Line voltage = $\sqrt{3}\ V_{PH}$ | Line voltage = $V_{PH}$ |
| Total power is = $3 \times V_{PH} \times I_{PH} \cos Ø$ | Total power is = $3 \times V_{PH} \times I_{PH} \cos Ø$ |
| Low insulation is sufficient. | High insulation is required. |

**74. What is stepper motor? (L1)**

Steeper motors are special machines known as brushless DC motors. These motors have special capability of dividing a full rotation into a number of equal steps. Input DC is connected across the voltage terminals; it converts the input pulses into a precisely defined increment in the shaft position with a fixed angle.

**75. What is Swinburne's method? (L1)**

Swinburne's method is used to determine the efficiency of machine in no load as well as pre-loaded condition. In this test a machine can be run as a motor or a generator. This method of testing can only be done in shunt and compound motors.

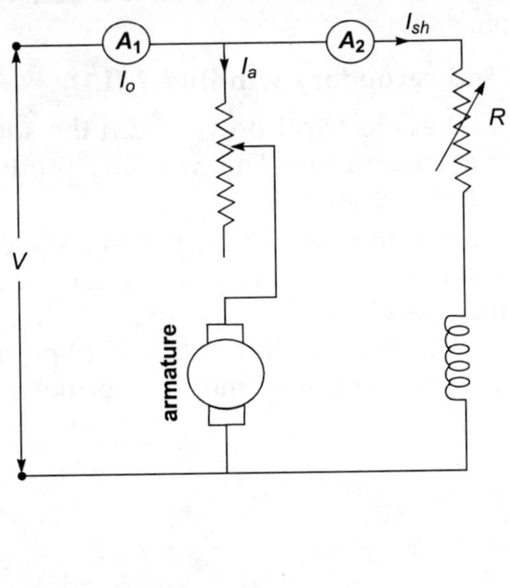

**76. What is the use of breather in a transformer?**

It is used to arrest the flow of moisture when the outside moisture enters the transformer.

**77. What is transformer and explain its working principle. (L1)**

Transformer is an electrical apparatus which is used to increase or decrease the voltage of an altering current.

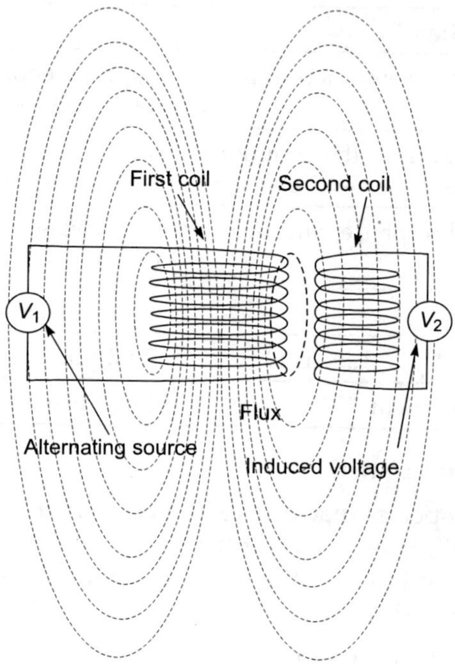


It works on the principle of the Faraday's law of electromagnetic induction, which states that the "rate of change of flux linkage with respect to time is directly proportional to the induced EMF in a conductor or coil". Hence, power is transferred by mutual induction.

**78. What are primary and secondary windings? (L1)**

The winding which takes electrical power from the source is generally known as primary winding of a transformer. This winding produces magnetic flux when it is connected to the electrical source.

The winding which gives the desired output voltage due to mutual induction in the transformer is commonly known as secondary winding of a transformer. In this winding, the flux, produced by primary winding, passes through the core, and links with the secondary winding. This winding is also wound on the same core and gives the desired output of the transformer.

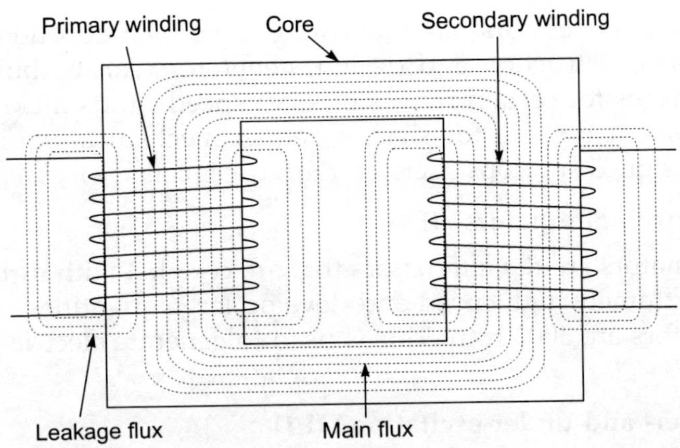


**79. What is step-up and step-down transformer? (L1)**

On a step-up transformer, there are more turns on the secondary coil than the primary coil. The induced voltage across the secondary coil is greater than the applied voltage across the primary coil or in other words the voltage has been "stepped-up".

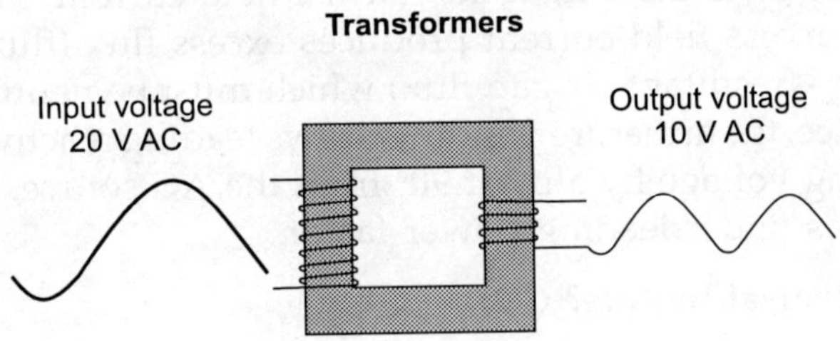


A step-down transformer has less turns on the secondary coil than the primary coil. The induced voltage across the secondary coil is less than the applied voltage across the primary coil or in other words the voltage is "stepped-down".

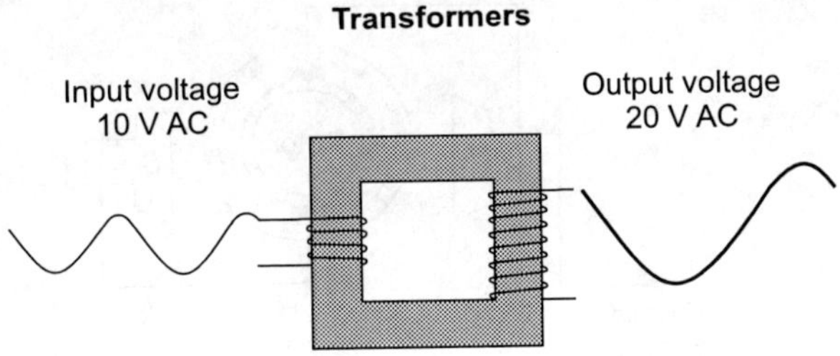


**80. What is tachogenerators? (L1)**

An electromechanical generator is a device capable of producing electrical power from mechanical energy, usually the turning of a shaft. When not connected to a load resistance, generators generate voltage roughly proportional to the shaft speed. With precise construction and design, generators can be built sto produce very precise voltages for certain ranges of shaft speeds, thus making them well suited as measurement devices for shaft speed in mechanical equipment. A generator specially designed and constructed for this use is called a *tachometer* or *tachogenerator*.

**81. What are turbo alternators? (L1)**

Turbo alternators are the alternators that are coupled with turbines. They basically run with extremely high speed and develop large quantities of power. The speeds of such motors are 3000 rpm. This is used to generate electric power directly from turbines.

**82. What is over- and under-excitation? (L1)**

In case of synchronous motors, if the field current is not sufficient enough to produce the required air gap flux as demanded by V, additional magnetizing current or lagging reactive VA is drawn from the AC source. This magnetizing current produces the deficient flux (constant flux-flux set-up by DC supply rotor winding). Hence, in this case the motor is said to operate under lagging power factor and is said to be under-excited.

If the field current is more than the normal field current, motor is said to be over-excited. This excess field current produces excess flux (flux set up by DC supply rotor winding – resultant air gap flux) which must be neutralized by the armature winding. Hence, the armature winding draws leading reactive VA or demagnetizing current leading voltage by almost 90° from the AC source. Hence, in this case the motor operates under leading power factor.

**83. What are universal motors? (L2)**

A universal motor is a special type of motor which is designed to run on either DC or single phase AC supply. The speed of the universal motors is in the range of 3500 rpm. These motors exhibit good efficiency in DC than AC which has reactance voltage drop.

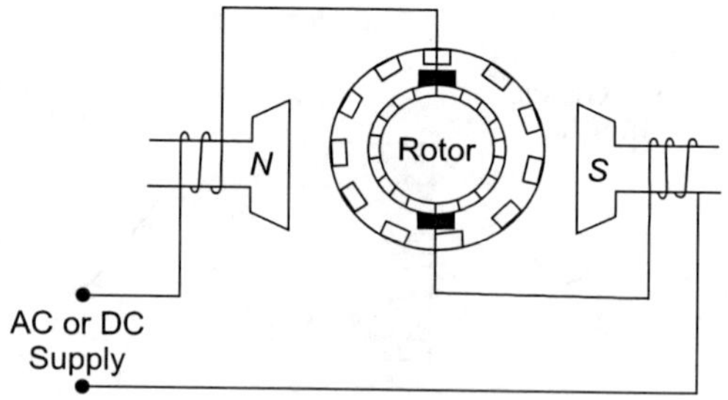


For example: Portable drills, cleaners, and drink and food mixers, sewing machines.

**84. What is ventilating ducts? (L2)**

In order to reduce the heat inside a transformer, ventilating ducts are provided. The cool air is continuously blasted inside the transformer by a fan. Thus, the air is filtered and enters the winding chamber via ventilating ducts.

**85. What is Ward – Leonard system? (L2)**

This is one of the speed control methods of DC motors introduced by Henry Ward in 1891. It is similar to that of armature speed control method. It is a basic armature control method. This control system consists of a DC motor $M_1$ and powered by a DC generator G. In this method the speed of the DC motor (M_1) is controlled by applying variable voltage across its armature. This variable voltage is obtained using a motor-generator set which consists of a motor $M_2$ (either AC or DC motor) directly coupled with the generator G. It is a widely used method of speed control of DC motor.

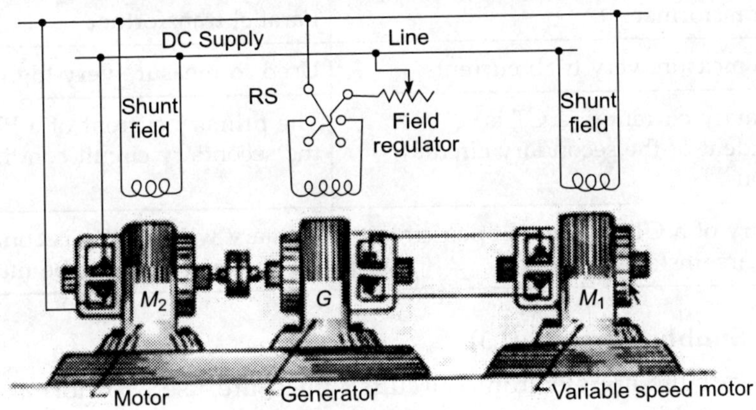


Ward Leonard system of speed control

**86. What is windage friction losses? (L2)**

All input energy is not available at the output due to losses. The windage friction loss is one such loss which takes place due to rotation of armature. It is proportional to the cube of speed. It is due to rotation of armature in air.

**87. What is an exciter? (L2)**

A generator magnetic field is energized by direct current. The DC field current is obtained from a separate source called exciter. Exciter is used in DC generator mounted on the shaft.

**88. Why do capacitors works on AC only? (L3)**

The reactance of a capacitor is given by $X_c = 1/2*Pi*f*C$,

In case of DC, the frequency is zero, hence reactance becomes infinite, and hence it blocks DC.

**89. Why does an inductor allow DC to pass? (L2)**

The inductive reactance is given by $X_l$ = 2* *Pi** *f** *L*. Hence, the value of reactance increases as there is an increase in frequency, but in case of DC, *F* = 0, hence the reactance is zero if *F* is substituted in the above formula. Hence, it allows DC.

**90. What is the working principle of circuit breakers? (L2)**

Circuit breakers are electrical devices which consist of fixed and moving contacts. They act as switching devices which are operated automatically/manually for controlling and protecting of electrical power systems. In case of any fault/abnormality in electrical equipment, these circuit breakers isolate the fault equipment preventing further damage.

**91. Comparison between CT and PT? (L2)**

| Current Transformer | Potential Transformer |
|---|---|
| Series transformer | Parallel transformer |
| Used to measure very high currents. | Used to measure very high voltage. |
| The primary current in a CT is independent of the secondary circuit conditions. | The primary current of a PT depends on the secondary circuit conditions. |
| Secondary of a CT cannot be open circuit on any circumstance. | Primary winding PT is connected across the line of voltage to be measured. |

**92. What is Snubber circuit? (L3)**

Snubber circuits are electronic circuits which are used to normalize the voltage transients, sudden surge, excess force, pressure, movement, etc. It is used to prevent a power system from external damage.

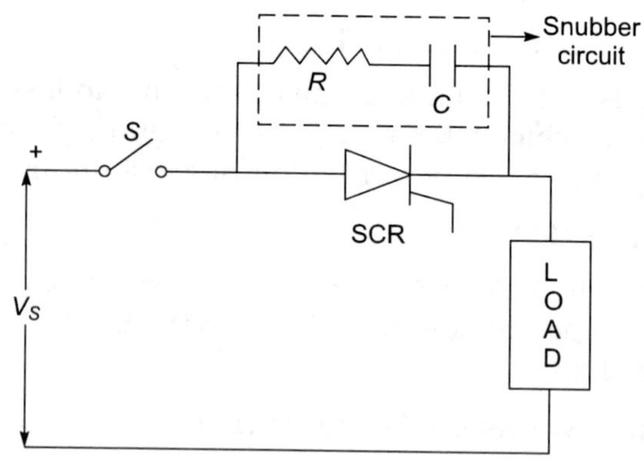

**93. Define name plate details of a machine. (L1)**

Name plate details are provided by the manufacturer stamped on a machine. Generally, they are used to understand the operational characteristics of a machine. Some of the important name plate details are:

- Voltage
- Frequency
- Phase
- Current
- Type
- Power factor
- kW or horsepower
- Full-load speed

**94. What is relay? (L1)**

It is an electromagnet which has a set of contacts which are operated electrically or mechanically. The main operation of a relay comes in places where only a low-power signal can be used to control a circuit. It is also used in places where only one signal can be used to control a lot of circuits.

Example: Telephone exchanges

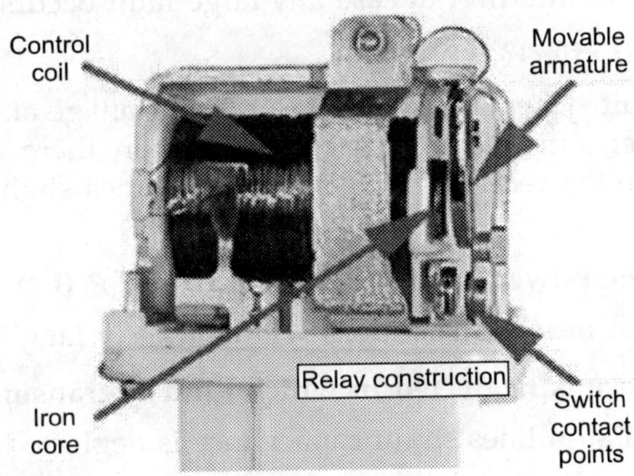


**95. What is VFD? (L1)**

VFD stands for variable frequency drives. It is power electronic drive which converts a basic fixed frequency, fixed voltage sine wave power (line power) to a variable frequency, variable output voltage, used to control speed of induction motor. It is used to control the speed of the three-phase induction motors.

**96. What are V and inverted V curves of synchronous motor? (L1)**

**V Curve** → Line current magnitude variation with respect to field current.

**Inverted V curve** → Variation of power factor with respect to field current.

**97. What would be the power factor of the transformer in no load and full load conditions? (L1)**

In no load → power factor is lagging.

In full load → Power factor is equal to power factor of the load that it is carrying.

**98. State the types of losses occurring in an induction motor. (L1)**

- Magnetic losses
- Electrical losses
- Mechanical losses

**99. What are lighting transformers? (L1)**

Transformers used in distribution system where no motors are connected is called lighting transformer.

**100. Why AC systems are preferred over DC systems? (L1)**

1. Alternating current is easy to maintain.
2. Operating cost of AC transmission is much lower than the equivalent DC transmission.
3. Power station generally produces AC; hence the conversion to DC can be avoided.
4. AC is easy to interpret in case any large fault occurs in the power network.

**101. What is Ferranti effect? (L2)**

In an open circuit system there will be a rise in voltage at the receiving end which is known as Ferranti effect. This happens when there is low load or no load condition where the receiving end voltage becomes slightly higher than sending end.

**102. What will be the power factor of industrial loads? (L2)**

The power factor of industrial loads will be mostly lagging.

**103. Shunt capacitance is neglected in which kind of transmission lines? (L2)**

In short transmission lines shunt capacitance is neglected.

**104. Which motor is suitable to drive fan? (L2)**

DC shunt motor is suitable to drive fan.

**105. What is LVDT? (L3)**

It stands for linear variable differential transformer. It converts displacement into corresponding voltage or current (electrical signals).

# 8 ELECTRICALS AND ELECTRONICS

Electrical and electronics engineering are also known as green areas which are required to learn and use in our day-to-day life Students should be motivated to learn these concepts. Following questions cover areas like basics of electrical and electronics and a few practical questions which answers why, where, why not kind of questions.

1. **Define Choke. (L1)**

   Choke is an inductor which creates high voltage needed for ionization of the gas.

2. **What is a damping system? (L1)**

   In measuring instruments, deflecting torque and controlling torque will be moving in opposite direction. This makes difficult for the pointer to come to rest. To bring the pointer to rest position damping torque is given to the system. Damping torque is proportional to the velocity of the moving system. This is known as damping system.

3. **What is the equation for torque? (L1)**

   $\tilde{I}$ = NBAI,
   T = Deflecting torque,
   B = flux density in air – gap,
   N = Number of turns in the coil,
   A = Effective coil area,
   I = Current in the moving torque.

4. **What is calibration? (L1)**

   Calibration is the process of finding the performance of a machine. In calibration the output characteristics of a device are practically obtained with respected to a master device. Hence, the uncertainty/variation between inputs to output is captured and plotted. Then suitable procedures are applied to adjust the uncertainty in the device.

5. **What is impedance matching? (L1)**

   To maximize the power transfer and to reduce the signal reflection in the load we must design the input impedance of a load in an electronic circuit. The process of selection of input impedance to that of the load impedance is known as impedance matching.

6. **Why is a starter used in tubelights? (L2)**

   A starter is a capacitor with stored energy inside. When the tube is powered on, the starter supplies a very high spark of current discharge which ionizes the gas inside the tubelight to make fluorescence effect for glowing tubelight in less time.

7. **What is the sequential logic circuits? (L1)**

   In few circuits they depend only on the present state of the input to give output but in a few circuit they also depend on the past state of the input and present state to give output. Such circuits which depend on the past state are known as sequential logic circuits.

   For example, data transfer mechanism use this sequential logic circuits.

8. **What is a flip-flop? (L1)**

   Flip-flops are the ones which help in storing the past output of the system in sequential logic circuit. In simple words feedback/memory element. The structure of the flip-flop is shown below.

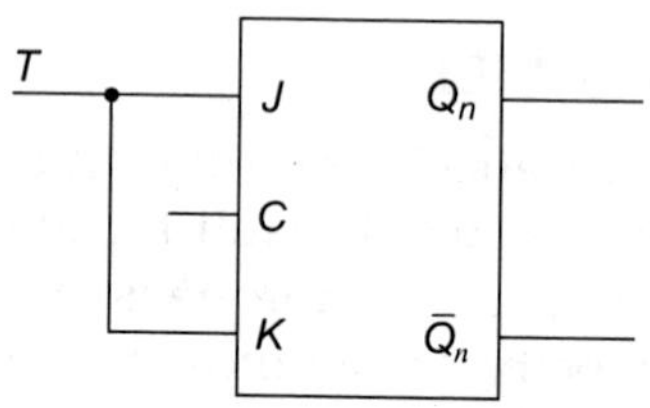


   It has

   - Two inputs
   - One clock generator
   - Two output.

9. **What are the different types of flip-flop? (L1)**

   The different types of flip-flop are SR, JK, D and T.

10. **What is a latch? How does it differ from flip flop? (L2)**

    Latch is also like flip-flop, which is based on past input state to give output. In flip-flop we have clock pulses so whenever there is change in the clock pulse it changes the output accordingly but in latch it checks input and changes output whenever there is change in input. There is no clock pulse based change in latch. So we tell latch is level triggered and flip-flop is edge triggered.

    Examples for flip-flop are registers, counters, etc.

    Example for latch is multiplexers.

11. **What is a magnetron?**

    It is an electron tube (cathode ray) used to generate or amplify short microwaves. It is a high powered vacuum tube that acts as a self-excited microwave oscillator. They are used in microwave owen for microwave generation.

12. **When a stationary charge is present and an electric charge moving at constant velocity, will there be radiation or not? (L2)**

    There will not be any radiation.

13. **In which flip-flops does race round condition occurs? (L2)**

    J K flip-flop.

14. **What is difference between a synchronous and an asynchronous counter? (L2)**

    Clock to all flip-flops are same in synchronous but in asynchronous counter the previous output is the next stage clock.

15. **How many 2 : 1 multiplexers do we need to construct a 8:1 multiplexer? (L2)**

    We need seven 2 : 1 multiplexers to construct a 8:1 multiplexer.

16. **What operation do you perform to get inverted binary bits? (L2)**

    One's compliment of each bit in a set of bits.

17. **Which logic gates are known as universal logic gates? (L2)**

    NAND and NOR.

18. **What is relation between VEM and K-map? (L3)**

    VEM is an extended version of K-map, K-map uses only 1's and 0's but in VEM we cannot only use 1's and 0's but an extra term called don't care term which is represented as '$x$'.

19. **How many half adders and how many XOR gates do we require to construct a 2 bit adder? (L3)**

    4 half adders and 2 XOR gates.

20. **What is a magnitude comparator? (L2)**

    It's a combinational circuit that compares two numbers and determines their relative magnitude.

    Example: take two numbers $A$ and $B$

    The magnitude comparator output will $A < B$ or $A > B$ or $A = B$.

21. **How to convert a moving coil into an ammeter? (L2)**

    By connecting shunt across the moving coil it gets converted to ammeter.

22. **What is AVO meter? (L2)**

    An ammeter, voltmeter, Ohm meter combination is known as AVO meter. Ammeter is used to measure current, voltmeter is used to measure voltage, Ohm meter is used to measure resistance of the conductor in ohms.

**23. What is Megger? (L3)**

Megger instrument is used for the measurement of resistance in the order of kilo or megaohms.

**24. What is mean by look angle? (L3)**

The elevation angle of satellite is known as look angle.

**25. What is tripping? (L2)**

Switching off a motor or an amplifier is known as tripping.

**26. What is the difference between RMS value and true RMS value? (L2)**

In true RMS we can get values for both sin and non-sinusoidal waveforms. But in RMS we can get values only for sinusoidal waveforms.

**27. What is Balun? (L3)**

It is a coupling transformer used to match impedance of the device.

**28. What is keying? (L3)**

Keying is a process where the radiation is limited by inserting a transmit switch to a transmitter.

**29. What is buckling? (L3)**

The battery should not be charged or discharged more than 25 Amps. If this happens the cell plates get bend which is known as buckling.

**30. How do we denote the tolerance of capacitor? (L2)**

It is denoted as '+' percentage. It is the permissible change in the capacitance value.

**31. What is the difference between tunnel diode and *p-n* junction diode? (L2)**

The doping level at *p* and *n* sides are heavy in tunnel diode while in *p-n* junction it is normal. Tunnel diode shows negative resistance but *p-n* junction diode does not show negative resistance.

**32. What is SCR? (L1)**

It is known as silicon controlled rectifier. It has four layers PNPN and three junctions anode, cathode and gate. It conducts only when it is forward biased. It is a power controlled device used in battery chargers, motor control, etc.

**33. What is hard switching of thyristor? (L1)**

The gate current is several times higher than the required gate current. The time taken to turn on the thyristor is reduced by enhancing the ratio of current with respect to time.

**34. What is TRIAC? (L1)**

TRIAC is the combination of two silicon controlled rectifiers which conducts in forward biased or reverse biased conditions. It is used for speed control of motor, phase control, etc.

**35. What is DIAC? (L1)**

It is a two-terminal semiconductor devices which is not used for controlling purpose but used for triggering purpose. It is used in light dimming circuits.

**36. What is the difference between LCD and LED? (L1)**

LCD is liquid crystal display which modulates the light and requires an external or internal light source. LED is light emitting diode which emits light during forward biased condition. It is capable of generating its own light.

**37. What is the advantage of using emitter resistance in the context of biasing? (L3)**

Emitter resistance increases stability by negative feedback.

**38. What is piezoelectric effect? (L2)**

Vibrations occur by stress over crystal or any other material which produces voltage that is directly proportional to the stress applied.

**39. Among all configurations which offers high output resistance? (L3)**

The common base configuration offers high output resistance.

**40. Among all configurations which offers lowest voltage gain? (L3)**

The common collector offers lowest voltage gain.

**41. Why an NPN transistor has higher frequency response than PNP transistor? (L3)**

The capacitive effect is less in NPN transistor due to the mobility of electrons. Hence, it has high frequency response.

**42. What is SOA? (L3)**

It is the safe operation area in the power device with proper current and voltage device.

**43. What is the other name of cycloconverters? (L2)**

The other name of cycloconverters is frequency changer. It converts one input frequency to other output frequency.

**44. What is the phase difference between inductive load and capacitive load? (L2)**

90 degree is the difference between inductive and capacitive load.

**45. What is masking? (L1)**

Electrically, when we suppress a signal then it is known as masking.

**46. What is meant by damping torque? (L2)**

The torque which stops the oscillation of the pointer in an instrument.

**47. What are the methods of developing damping torque? (L2)**

Air friction method
Fluid friction method
Eddy current method.

**48. How can we convert a moving coil meter into a voltmeter? (L2)**

By connecting a high resistance in series with coil we can connect moving coil into voltmeter.

**49. What is the use of moving iron instrument? (L2)**

It helps measure high amount of current (both AC and DC).

**50. What is an electrostatic voltmeter? (L1)**

It is a measuring instrument which is designed to measure high voltage which works on electrostatic attraction.

**51. What is a parametric amplifier? (L2)**

At microwave frequencies the reactance of the device is varied in time by alternating voltage which is known as pump voltage. When a suitable relation is maintained between the pump voltage and the frequency of the signal amplified, then pump energy is transferred to the signal.

**52. What is ZigBee?**

ZigBee is a low-cost, low-power, wireless mesh networking standard. First, the low cost allows the technology to be widely deployed in wireless control and monitoring applications. Second, the low power usage allows longer life with smaller batteries. Third, the mesh networking provides high reliability and more extensive range. ZigBee relies on the basic 802.15.4 standard to establish radio performance

**53. What is a plasma accelerator? L4**

A high velocity jet of plasma is formed by an electric arc, travelling wave, etc. This is known as plasma accelerator.

**54. What does friction compensation do? (L3)**

It adjusts the shading bands in energy meter.

**55. Differentiate between a rheostat and a potentiometer? (L2)**

A potentiometer measures small resistance whereas a rheostat measures high resistance.

# 9 ANTENNAS AND RADARS

Antennas and radars play a vital role in communication which should be known to students. This part covers basics of antennas, design and their types, mathematical analysis, representation and so on.

**1. What are microstrip antennas? (L1)**

It is also called pitch antenna. A microstrip patch antenna is made up of a conducting patch of any planar or nonplanar geometry on one side of a dielectric substrate with a ground plane on other side. It is used to transmit or receive electromagnetic waves.

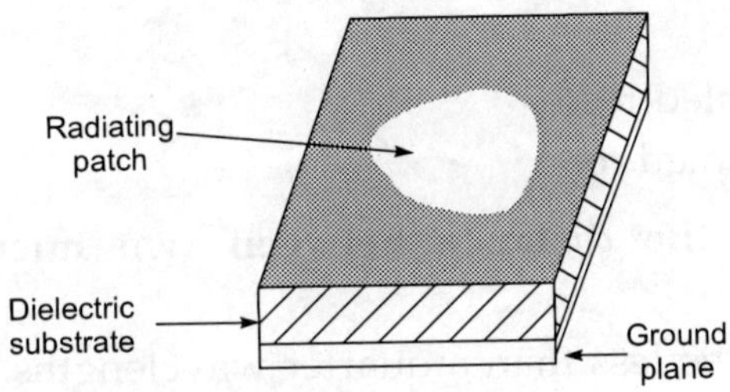


**2. What are the applications of microstrip antennas? (L1)**

(i) Mobile and satellite communication application
(ii) Global positioning system applications
(iii) Telemedicine application
(iv) Radio frequency identification
(v) Rectenna application
(vi) Worldwide interoperability for microwave access
(vii) Radar application

**3. What are the advantages and disadvantages of a microstrip? (L1)**

| Advantages | Disadvantages |
|---|---|
| Moderately efficient | Low efficiency |
| Low power | Low gain |
| Thin profile and hence less weight | Large ohmic losses |
| No cavity backing | Low power handling capability |

**4. Define antenna feed. (L1)**

Feeding is most difficult part of any antenna system. Antenna feed is a component in antenna which feeds the radio waves to the rest of the antenna structure. It acts as an interface which collects the incoming radio waves, converts them into electric currents and transmits them back to the receiver.

**5. What are the 2 types of antenna feeding? (L1)**

- Non-contacting feed
- Contacting feed

**6. What are contacting and non-contacting feeding? (L1)**

In contacting feeding a patch is directly fed with RF power using the contacting elements such as microstrip line or coaxial line.

In non-contacting feeding the patch is directly fed with RF power using the contacting element such as microstrip line or coaxial line.

**7. What are the feeding techniques used in antennas? (L2)**

1. Microstrip line feed
2. Inset feed
3. Co-axial feed
4. Aperture coupled feed
5. Proximity coupled feed

**8. What should be the distance between two microtrip lines for them to couple? (L2)**

The distance must be less than a quarter wavelengths.

$D < \lambda/4$

**9. What is MUF? (L2)**

MUF is maximum usable frequency. It is the highest radio frequency which can be used for transmission between two points via reflection from the ionosphere at a specified time, independent of transmitted power.

**10. What are the three main functions of antenna grounding? (L2)**

Grounding of an antenna is essential as it provides the below mentioned functions.

Electrical safety

Stray RF suppression (or simply RF grounding)

Lightning protection

**11. Why are Smith charts? (L2)**

Smith charts are used in RF engineering to solve the problems in transmissions lines and matching circuits. It is also used to represent the RF parameters and behaviour of transmission lines.

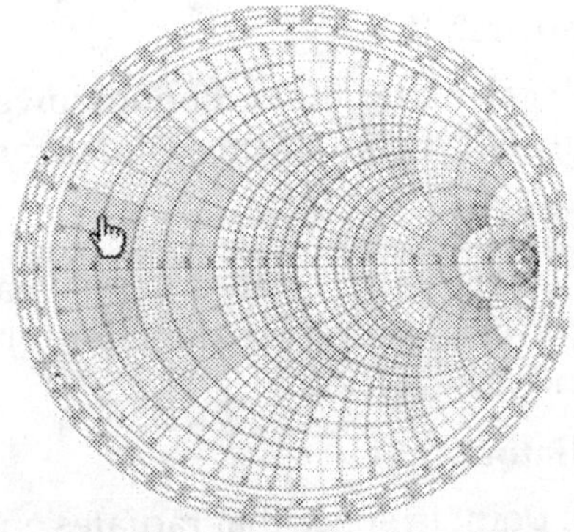

12. **What is critical angle? (L1)**

The critical angle is defined as the angle of incidence that provides an angle of refraction of 90 degrees. When light is incident on a material at its critical angle light moves along the surface of the material.

13. **What is directional antenna? (L1)**

It is also called a beam antenna. It is a RF wireless antenna designed to function more effectively in some directions than in others. The purpose of that directionality is improving transmission and reception of communications and reducing interference. Depending on the half power beam width (HPBW) an antenna is determined as directional antenna.

14. **What is the relation between HPBW and BWFN? (L3)**

The half power beam width (HPBW) is the angular separation in which the magnitude of the radiation pattern decreases by half (or –3 dB) from the peak of the main beam. BWFN stands for beam width, first nulls. This is the angular separation from which the magnitude of the radiation pattern decreases to zero (negative infinity dB) away from the main beam. HPBW is calculated as BWFN/2.

15. **What are the other ways to find the directivity when the gain of the antenna is not given? (L2)**

When the gain of the antenna is not given we can find the directivity using average radiation intensity.

16. **What are the antenna pattern parameters? (L2)**

- Radiation lobe
- Main lobe
- Minor lobe
- Side lobe
- Back lobe
- Half power beam width (HPBW)
- First null beam width (FNBW)

**17. What is effective aperture? (L1)**

It is a parameter which calculates the receiving power of an antenna in the effective area or effective aperture. It is an effective area from which power is absorbed.

**18. What is antenna gain? (L1)**

Antenna gain is relative measure of an antenna's ability to direct or concentrate radio frequency energy in a particular direction or pattern. The measurement is typically measured in decibels (dbi).

**19. What is isotropic radiator? (L1)**

It is a theoretical single point in space that radiates energy equally in every direction similar to the sun radiating its light. The isotropic radiator exhibits the same magnitude or properties when measured in all directions. It has no preferred direction of radiation. It radiates uniformly in all directions over a sphere centred on the source.

**20. What is RTD? (L2)**

RTD stands for resistance temperature detector. These sensors are used to measure temperature. They are used in industrial applications to measure the temperature less than 600 degrees. They are made up of nickel or copper material with a fine wire wrapped around a ceramic or glass.

**21. Explain the solenoid valve? (L1)**

It is an electrically operated valve. It consists of an electromagnet in which a magnetic plunger moves. The plunger is connected to a switch that tends to open or close. There are 2 types of solenoid valves.

- Normally open
- Normally closed

They are used in safety doors in industries.

**22. What is Radar? (L1)**

Radar is defined as radio detection and ranging which helps in seeing objects/ detects objects which are not visible, e.g., objects during night or while travelling in clouds, etc. It transmits and detects reflected electromagnetic waves. It gives details about the location (longitude and latitude), size, velocity of the object, etc.

**23. What is DMC? (L2)**

DMC is stated as digital magnetic compass. It gives latitude and longitude of the object instead of going for difficult/complex algorithm.

24. **What is phased array radar? (L1)**

    Phased array radar tracks the object the object without changing the direction of the radar. In simple words it helps in navigation.

25. **Which antenna gives a donut shape radiating pattern? (L3)**

    Dipole antenna.

26. **What part of the radiation lobes we will have maximum power? (L3)**

    Main lobe of antenna.

27. **When electric field is in constant direction what kind of polarization does it will have? (L3)**

    Linear polarization.

28. **What are the patterns of lambda dipole and lambda/2 dipole and in what way do they differ? (L3)**

    Lambda dipole and lambda/2 dipole have same kind of radiation pattern but the directivity of lambda dipole is more compared to lambda/2 dipole antenna.

29. **Which shape of a wire will radiate? (L2)**

    Circular, square and triangle shaped conductors will radiate but straight conductor will not radiate.

30. **What should be the antenna output impedance to radiate efficiently into free space? (L2)**

    Antenna output impedance should be equal to free space impedance.

31. **What is the value of magnetic flux in a closed surface? (L2)**

    ZERO.

32. **What is the kind of code used in K-map? (L2)**

    Gray code.

# 10 MECHANICAL AND ELECTRONICS

This part gives a good idea about recent trends and research in mechanical and electrical areas which is helpful for the students for their enhancement of knowledge.

1. **What is MEMS? (L1)**

   MEMS is stated as micro electromechanical systems. It is a technology that is used to fabricate gears, beams (mechanical elements) with capacitors, inductors, etc. In simple words, it integrates active and passive elements.

2. **What is etching? (L1)**

   It is the process of applying chemical or electrical effects to remove a part of layer for some functional purpose.

3. **What is the difference between lithography and etching? (L1)**

   Etching removes a part of layer whereas lithography does not remove layer but just patterns the material structure.

4. **What is nanotechnology? (L1)**

   It is used for miniaturization process to give more resource consciousness and efficiency. It is used in household applications, automotive industries, etc.

5. **What are shape memory alloys? (L1)**

   These are the alloys which exist on both high and low temperatures when heated.

6. **What is a microfluid system? (L2)**

   Micro devices which are used to transport fluid and store it are known as micro fluid system.

7. **What is HEMT? (L3)**

   It is stated as high electron mobility transistor. It is used in high frequency amplifiers.

8. **What is the difference between nonlinear model and 'S' parameter of a transistor? (L2)**

   Nonlinear model is a circuit model of a transistor with which we can choose our own BIAS point whereas in 'S' parameter BIAS is fixed.

9. **What is metamaterial? (L3)**

   It is man-made material which deviates electromagnetic waves instead of reflecting them. It is used in an electromagnetic spectrum to make any given substance invisible.

# 11 COMPUTER NETWORKS

This chapter deals with the area of computer networking. All are very important and basic questions taken out from computer networking concepts. Readers can practise some of these questions in parallel, practically with the answers provided.

1. **What is a network? (L2)**

   A network, predominantly a computer network is a collection of hardware components and computers interconnected by communication channels governed by network protocols. The aim of computer networks is to facilitate sharing information and resources through various means with the help of technology; from multiple sources to multiple destinations. Networks may be wired or wireless; and it can comprise various components, for instance, personal computer, mobile phones, personal digital assistant, printers, hubs, routers, etc.

2. **Why do we need computer networking? (L2)**

   To share information, one needs to be connected to a network. With the vast amount of data generated every day, it can be meaningful only if it is transmitted accurately and proportionately. We cannot imagine a world without passing information. One common example that would make sense is Facebook or Twitter. It is referred to as social networking sites because it helps in sharing information to a group or public. So basically, the need of a network is to share information.

3. **What is a topology and what are the topologies commonly followed in the computer networking? (L2)**

   A topology is nothing but network's virtual shape or infrastructure. It is the means of integrating all the nodes in a network in a particular fashion. Network protocols, configuration and network languages are used to dispatch information between the nodes. A topology is simply a schematic description of the arrangement of a network, including its nodes and connecting lines.

   Commonly followed topologies are:

   - Bus topology
   - Ring topology

- Star topology
- Tree topology
- Mesh topology.

Reader can refer to the following figure to get an idea of how the topology would look like.

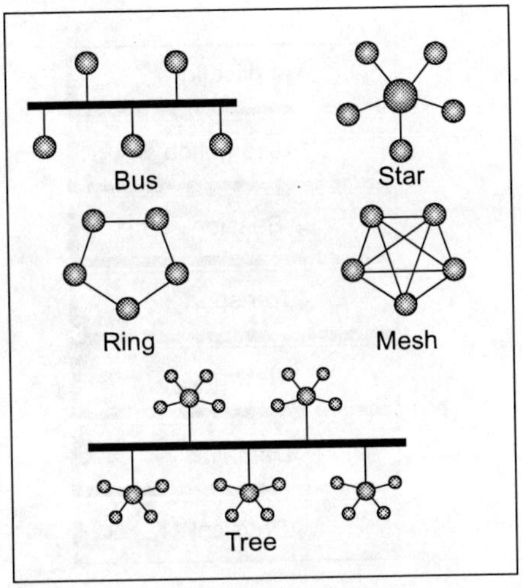


**Network topologies**

4. **Define a link with respect to networking. (L1)**

When two devices are said to be connected, the link refers to the connectivity. Normally, one would refer to cable used to connect two devices as link in a wired network.

5. **What is a node? Define it. (L1)**

Anode normally represents the point where connection takes place. Any device or equipment which is part of the network can be referred as a node.

6. **What is OSI and what was need for it?**

OSI stands for open system interconnection. When the computers were first connected in a network, moving the information could not be done correctly and it resulted in confusion. Therefore, a framework needed to be evolved that could serve interoperability between network channels, protect the integrity of information; alongside hiding the underlying structure to avoid ambiguity. Hence, International Standards Organization (ISO) created the OSI framework in early 1980s and it is being followed till date. OSI is a standard networking model that uses divide and conquer approach, wherein one big problem is spilt and seen as 7 different layers.

7. **Who framed OSI modeling? (L1)**

International Standards Organization (ISO).

8. **How is OSI structured and what are the layers in it? (L2)**

OSI layering basically uses divide and conquer approach wherein one big problem is spilt and seen as 7 different layers. OSI model is shown in the following figure. People remember the layers by the following clue "Please Do Not Touch Steve's Pet Alligator" where P represents the physical layer, D represents the data-link layer and so on.

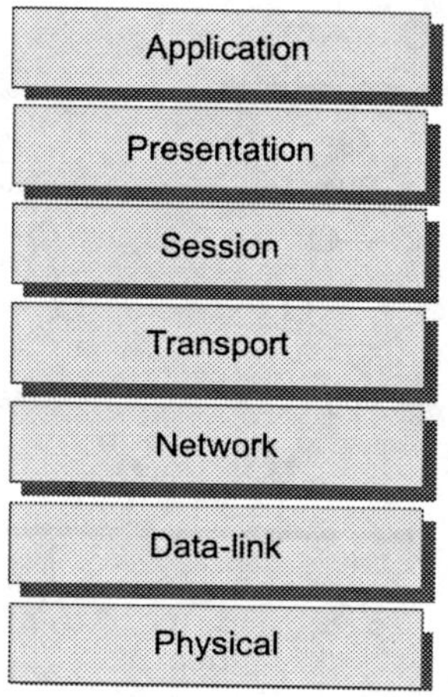


**Layers in OSI.**

9. **One can access files in public servers without having to worry about the identity. This is achievable through a protocol; what is it? (L1)**

One can use anonymous FTP to gain access to a server. In this, there would not be any need for the users to reveal their identity, instead they are permitted to go on with the guest login to gain access for the files in the server. But, this is not preferred when the data to be accessed is of high importance. If the data is of high value and confidentiality, then, SFTP (Secure FTP) is preferred.

10. **What is a repeater and its function? (L2)**

Repeaters are like re-generator devices, placed in a network for the propagation of signals. It receives a digital signal from a transmitter, amplifies it and then will re-transmit with better strength. When the signal has to be sent for a longer distance, the repeater will be very much useful. It is a layer –1 device. A simple diagrammatic representation is shown below.

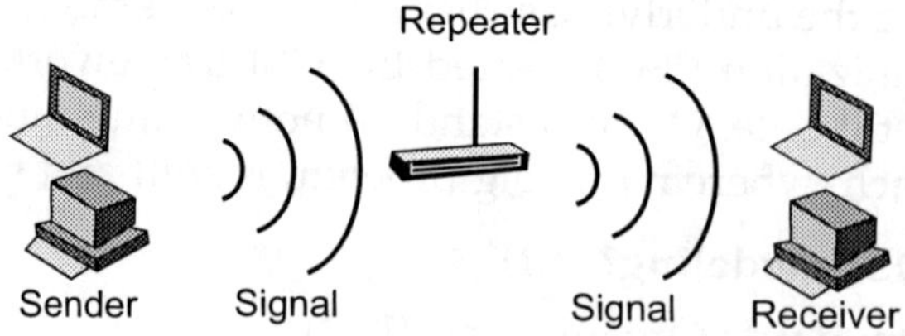


**How repeater works?**

**11. What is the use of a hub? Which layer does it work? (L2)**

Hub is deployed to facilitate data sharing between computers in an easy manner within a network. Whenever star topology or any other local network topology is used, the hub becomes inevitable. Hub acts as a major connection point; having many ports. It is designed to work as a multiport repeater indeed. Hubs operate widely in the physical layer. A simple deployment example is shown below.

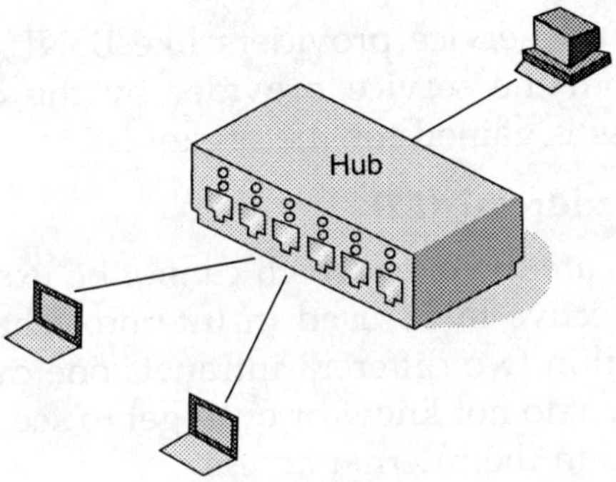


**Working of a repeater**

**12. How is router useful in computer networking and where it is located in OSI layer? (L2)**

A router runs for any information to be transmitted from one network to another. Controlled by network protocols like TCP/IP, a router divides the network segments as logical parts called subnets. Routers communicate in an intelligent way by transferring the packet from source to destination with their packets headers, addresses and subnet details; thus avoiding congestion in a short route. It is the most vital component working in layer – 3 of OSI mode. Figure below represents the way the router works.

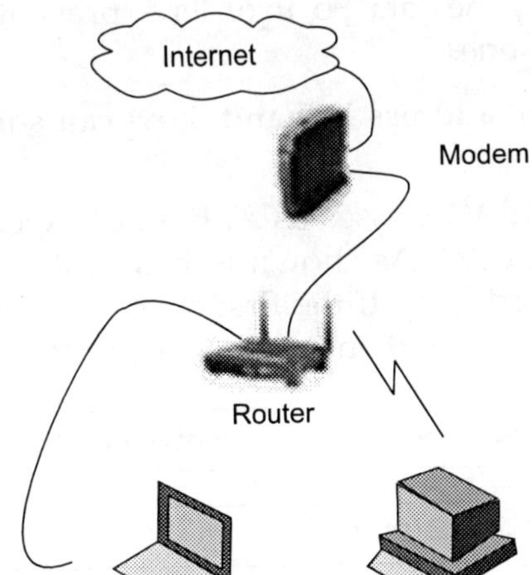


**Working of a router**

**13. Define the roll of NAT (Network Address Translation) in the computer networking. (L2)**

There could be multiple computers used in a particular network. For an instance take a university, there should be 1000s of computer available. They all will be sharing the same internet connection. This is made possible through NAT.

**14. What is public IP address? (L2)**

We all know there are service providers like BSNL, Airtel. Public IP is the one obtained/leased from the service provider by the organization through which direct internet access is gained for the organization.

**15. What is private IP address? (L2)**

Private IP addresses are the ones which cannot be used for connecting to internet, i.e., they are not effective to be used in Internet. They are predominantly useful on the intranets. Within two different intranets one can use the same IP, i.e., same private IP as they two do not know or even get to see other. One would need NAT or proxy server to gain the internet access.

**16. How is static IP addressing different from dynamic IP addressing? (L2)**

When static IP addressing is preferred, following points are to be noted.

- Static IP addressing is done manually. Secondly, the IP assigned statically will not be changed and it will be used always by the machine.
- If dynamically assigned, the IP address will get changed periodically based on which network the system is connected in. Dynamic assignment is not done manually and done through the protocol called DHCP (Dynamic Host Configuration Protocol).

**17. Why the cables used for networking are shielded? (L2)**

It is to prevent the noise which can get mixed to the data. This noise has to be prevented from entering the data. So, in order to prevent unnecessary noise entering the data, shielding is done.

**18. Given with an IPV4 IP address in hand, how can someone identify the class it belongs to? (L2)**

Most frequently used IP V4 classes as A, B and C. One can classify an IP as A, B or C based on the first octet. As shown in below, if the first octet starts with 0, it is classified as Class A address. If the first octet starts with 10 then it comes under Class B. If it begins with 110, then it falls under Class C IP address.

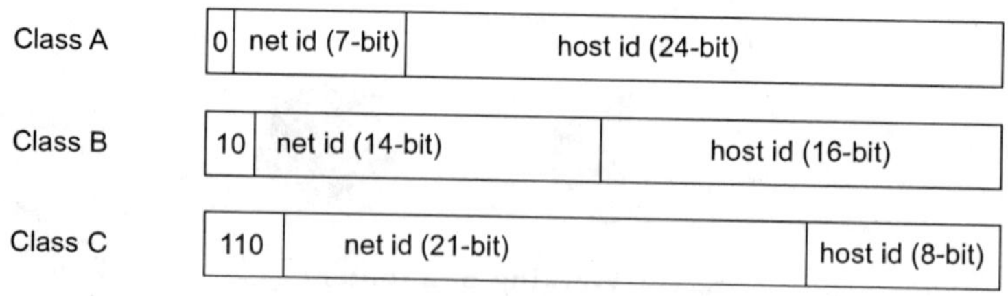

**19. What is TCP and how is it useful? (L2)**

TCP is known as transmission control protocol, which is the most commonly used protocol in the Internet. Excellent error correction guarantee and "guaranteed delivery" of data is possible with TCP. This is connection oriented protocol and the data packets will be moved from source to destination only after the connection is established. Flow control is provided with TCP. If there is a corruption in the data is sent, TCP ensures retransmission thereby ensuring guaranteed delivery. One can understand this scenario from the figure below.

**TCP – Connection oriented protocol**

**20. What is UDP How is it different from TCP? (L2)**

Unlike TCP which guarantees data delivery and which is a connection based protocol, with UDP there is no guarantee that the data will be delivered completely to the destination. This is a connectionless protocol. UDP does support port numbers and checksums on its datagrams. Wherever speed is the requirement, there UDP can be administered. Streaming video, etc., are best applications where UDP is employed, whereas applications like banking website (where security and guaranteed delivery is important) and all are example for TCP. One can understand the scenario from figure.

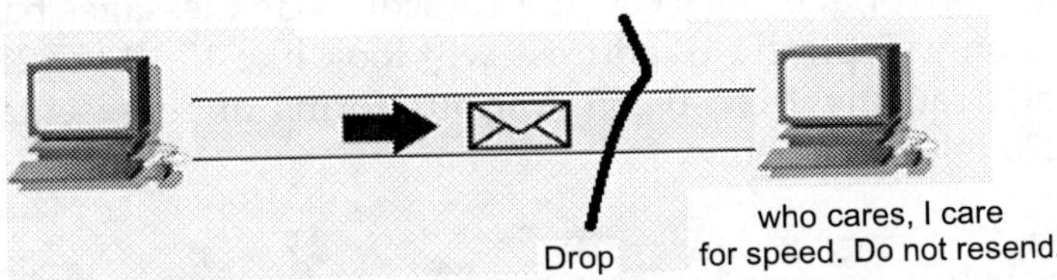


**UDP – Connection-less protocol**

**21. Which one would you prefer as a user UDP or TCP? Justify. (L2)**

It depends on the situation. In TCP, a connection will be established first and will be retained until the data transfer is over. So reliability is ensured to a great extent in TCP. Guaranteed delivery is the key when TCP is used. UDP is used for the speedy delivery. There is no guarantee that the data will be delivered in full to the destination. But it does it so fast. Wherever speed is the need, UDP can get in. The reason for the speed is, there is no flow control or error correction. UDP is deployed widely in streaming applications.

**22. How is TCP better than UDP? (L2)**

In TCP delivery is guaranteed, since it establishes a dedicated connection. Hence, TCP scores better than UDP when it comes to reliability of data transfer.

**23. Compare TCP and UDP. (L3)**

| TCP | UDP |
|---|---|
| Connection oriented | Connectionless |
| Reliable | Unreliable |
| Flow control and retransmission | No flow-control |
| Guaranteed delivery | Best effort delivery |
| Acknowledgements available | No acknowledgments. |

**24. Firewalls are one of the most frequently used terms in computer networking world. What is it? How is it useful? (L2)**

Assume, some attacker wants to attack you when you are at home where you can protect yourself by having a firewall. Which means, the attacker would not be able to jump over the firewall and attack you as it will hurt him. Same is the case with a network. Firewalls can prevent the attackers/hackers from attacking your network. It can also prevent the users from accessing unauthorized sites. It will avert viruses from entering networks.

**25. What is an IP address? Explain (L2)**

An IP address is a unique logical identification for a node which is connected on a network. Networks using TCP/IP protocol route messages based on this unique IP address only. Typically IP address will look like 10.10.127.220. IP address is of 32 bits each ranging from 0-255. Various forms of representation of an IP, say 10.10.10.220 is shown below.

```
      10.                  10.                  10.                 220          => IP address as decimal.
128 64 32 16 8 4 2 1   128 64 32 16 8 4 2 1   128 64 32 16 8 4 2 1   128 64 32 16 8 4 2 1   => 8421 Representation.
 0  0  0  0 1 0 1 0     0  0  0  0 1 0 1 0     0  0  0  0 1 0 1 0     1  1  0  1 1 1 0 0    => IP Address as Binary.
```

**Various representations of IP.**

**26. What is the need of a MAC address if IP address is known? (L2)**

An IP address is for the unique identification of the network while MAC address is for the exact identification of the port where the communication is initiated. The IP address is sort of a logical address that is assigned to each device on the network, while the MAC address on the other hand is a physical address that is used for the network device or hardware. In simple terms, IP address is like the street address while MAC address is like the door number. Without the MAC address we will not be able to reach out to the exact port.

**27. Who is maintaining the Internet world? (L2)**

IANA, Internet Assigned Numbers Authority. They are responsible for internet to run smoothly. IANA maintains domain names, IP addresses and protocols. More technical information on IANA can be found out from www.iana.org

**28. How is MAC address framed? (L2)**

A MAC address lengths to 6 bytes, in which the first three bytes represent the vendor details which is issued by the manufacturer. These three bytes are called **OUI, Organizationally Unique Identifier**. Embedded into a network hardware (NIC) while it is manufactured, network switches or routers store this address and forward data only if the MAC address matches and has access control. The next three bytes are called UAA, universally administered address. Since all cards from a given manufacturer have the same first 3 bytes, the UAA is useful in making the MAC unique. Figure below represents the MAC format.

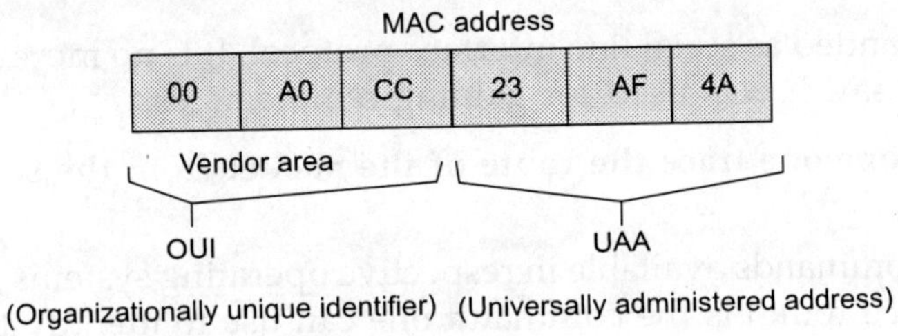


**MAC address.**

**29. What is Ethernet and what is its frame format?**

Ethernet is a local area network protocol introduced by Xerox Corporation. Ethernet is basically a standard that enables communication over a single cable, which is shared by all devices (computers/nodes) on the network and takes care of how the data can be transmitted to many devices simultaneously. Once any device gets attached to this cable, it then gains access to communicate to the other elements attached to the same cable. There are seven fields in the Ethernet. It does not have any provision for acknowledgement. The seven fields of the Ethernet are shown clearly in the figure.

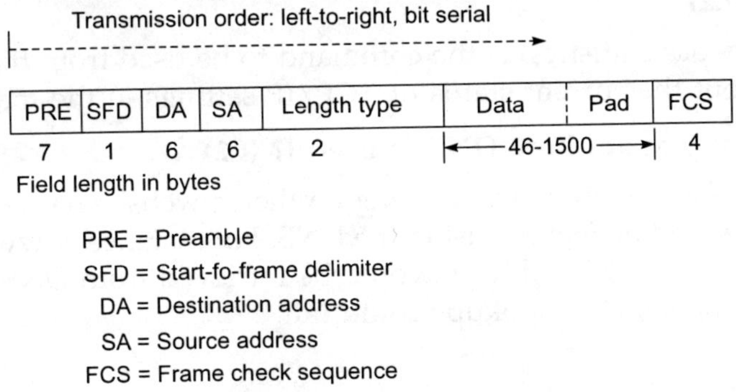


**Ethernet frame format.**

**30. What is IETF and expand RFC? (L2)**

The Internet Engineering Task Force (IETF) is an open global group of network designers, operators, sellers, and scientists concerned with the development of the Internet architecture and its flawless operation. It is open to any interested individual. RFC, request for command, is the repository maintained by IETF.

**31. How can someone define the roll of Gateway used in networks? (L1)**

One can easily understand that there are multiple networks coming together to get connectivity and to establish communication. When two different networks, not knowing each other communicate, the gateways would help in translation. Segment A can easily communicate to Segment B though they are different in all aspects with a Gateway in place. It is a translator basically.

**32. Other than TELNET, FTP, SFTP, there was another protocol in place called SLIP. What is it used for? (L1)**

SLIP is expanded as serial line interface protocol. It is no more in usage or existence one would say. It was used for gaining remote access.

**33. How can someone trace the route of the packet from the source to destination? (L2)**

There are commands available in respective operating systems for finding the route. Tracert (trace route) is the command one can use to identify the route taken from the source to destination by the packet in Windows platform.

**34. Dynamic assignment of IP address is made via protocol DHCP. What is it? Brief on the same. (L2)**

DHCP stands for dynamic host configuration protocol. It helps in assigning IP dynamically whenever a machine enters into a network. Also, the IP will not be assigned permanently, instead, it gets leased to the node connected to the network. Once, the system connected to the network is moved out, IP address will be taken back and the same address can be allotted to any other machine coming into the network, in future.

**35. How can someone identify or learn about the TCP/IP settings in a Windows machine? (L2)**

netstat (network statistics) is the command to be used from the command prompt to learn about the current status of TCP/IP settings in the machine.

**36. Briefly convey about how DNS is useful? (L2)**

DNS stands for domain naming system. When a website name is typed, equivalent IP address would be fetched using the DNS. For instance, www. google.co.in when typed, its equivalent IP address would be retrieved from DNS. When we misspell the website name, DNS lookup would fail.

**37. OSI can be seen as network support layers and user support layers. What are layers coming under the above split? (L2)**

OSI layers can be seen as two network support layers and user support layers. Out of the seven layers, the following three qualify as network support layers:

(a) Physical layer
(b) Data link layer
(c) Network layers

While the following three come under user support layers.

(a) Session layer
(b) Presentation layer
(c) Application layer

**38. How can someone get to know the MAC address in a Windows machine? (L2)**

Getmac is the command used to get the details of the MAC address in WINDOWS platform. A snapshot is presented below in figure with getmac being used.

```
C:\Users\shriramkv>getmac

Physical Address
===================
4C-0F-6E-FF-16-69
```

**getmac result snapshot.**

**39. How can someone get to know the details of IP addresses configured, in a Windows machine? (L2)**

One can use ipconfig to acquire all the IP configuration settings. Snapshot is presented below.

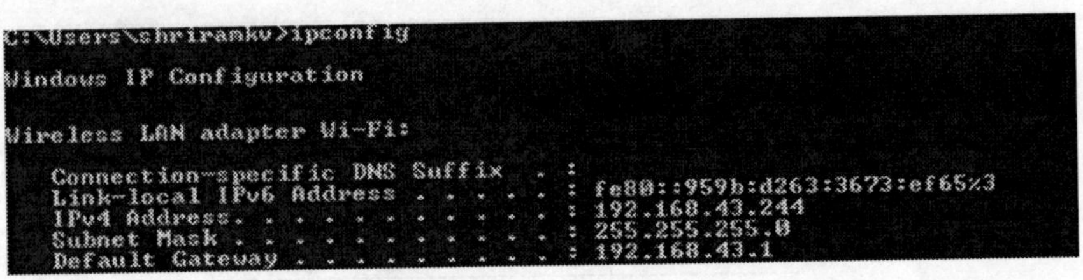

```
C:\Users\shriramkv>ipconfig

Windows IP Configuration

Wireless LAN adapter Wi-Fi:

   Connection-specific DNS Suffix  . :
   Link-local IPv6 Address . . . . . : fe80::959b:d263:3673:ef65%3
   IPv4 Address. . . . . . . . . . . : 192.168.43.244
   Subnet Mask . . . . . . . . . . . : 255.255.255.0
   Default Gateway . . . . . . . . . : 192.168.43.1
```

**ipconfig result snapshot.**

**40. How can someone get to know the details of IP addresses configured, in a Linux/ Unix machine? (L2)**

ifconfig is the command to be used in Linux to get the details of the IP configuration. Normally, ifconfig is used as ifconfig –a.

**41. What is NSLOOKUP command used for in Windows? (L2)**

nslookup helps to find out the name of a PC in a network. One may know IP address of a PC and would like to know the name of the PC. Nslookup will provide the info.

**nslookup<IP OF THE TARGET MACHINE>**

**42. When can someone call a network effective and efficient? (L2)**

Basic expectation from any network is performance, i.e., the network should be providing service all time without failing. If there is a failure, it should get fixed automatically without loss of data. The response time should be lesser. The network should be reliable. The network should be intrusion free. The network should filter the unauthorized access attempts. It should prevent the data from viruses. It should be scalable. This list can keep growing. To make it precise and short, a good quality of service is what a network should provide to be called an efficient and effective network.

**43. What is a bus? What is the basic principle behind ethernet? (L2)**

Bus can transfer data between components within a computer or between computers. It is a physical connection like a pathway that is required for communication through a single or multiple channels to network devices. The basic principle governing Ethernet is carrier sense multiple access with collision detection (CSMA/CD).

**44. What are the classes of IP address? (L2)**

There are 5 different classes of IP namely Class A, Class B, Class C, Class D and Class E. An IP is basically split into two halves. One is network ID and the second is host ID. Classes of IP are formulated based on this net ID and host ID. The various classes of IP are explained in figure.

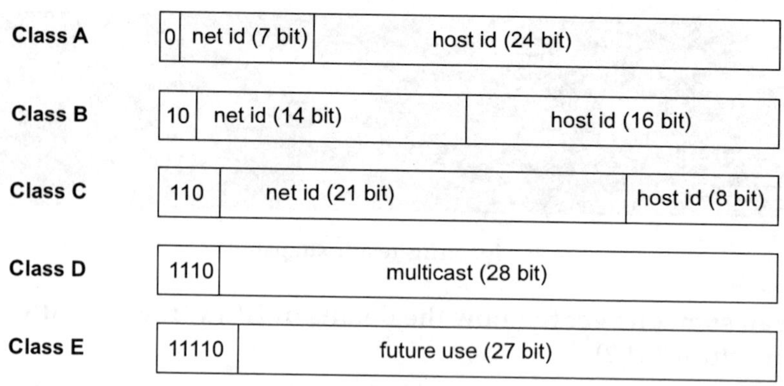


**Classes of IP.**

**45. Which class of IP address would you prefer if you have a huge number of nodes to be connected to your network? (L2)**

Class A. In a Class A - IP address, the first octet (8 bits) is the network ID (126 networks) and the last three octets (24 bits) correspond to the host ID. Hence, a total of (2 ^ 24)-2 IP addresses can be assigned under Class A.

**46. What is the device that can work in all the 7 layers? (L2)**

Gateway is the device which works in all the 7 layers.

**47. TELNET is a protocol used in one of the seven layers. Which is that layer? (L2)**

TELNET is an application layer protocol.

**48. What is the use of FTP? (L2)**

FTP is used for the file transfer between the terminals with the get and put commands.

**49. Expand – SMTP. (L2)**

It is simple mail transfer protocol. It governs the email communication happening between nodes in a networks.

**50. What is the use of SNMP? (L2)**

SNMP is used to manage devices on IP networks.

**51. There are certain factors which can influence the performance of a network. What are they? (L2)**

First, the number of users. If limited users are using a network, then the load, congestion, etc., would be limited and the network would provide good performance. Secondly, the infrastructure plays a major role, i.e., the type of the hardware and software quality does matter.

**52. Define bandwidth. (L1)**

Bandwidth is one of the most frequently used terms when someone talks about the performance of a network. Bandwidth defines "the number of bits transmitted over the network" in a specific period of time.

**53. Define latency. (L1)**

Latency specifies the amount of time needed for the packets (data) to reach the destination from the source, i.e., it defines the travel time. Lesser the latency, better the performance.

**54. Explain the terms (a) unicast, (b) multicast and (c) broadcast. (L1)**

When the data is sent from a source to only one recipient, it is referred as Unicasting. Example: Making a phone call to one specific number.

When the data is sent from a source to specific group of recipients out of all available, it is referred to as multicasting. Example: Circulars being sent to only managers from top management as an E-mail.

When the data is sent from a source to all the nodes in the network, it is referred as broadcasting. Example: TV telecast.

55. **Which layer in OSI has responsibility towards deciding or governing the data rate? (L1)**

It is physical layer's responsibility. Based on the hardware, channel and medium, etc., the data rate can be decided.

56. **Flow control and connection control are taken care of by one of the layers in OSI model. Which layer is that? (L1)**

All the connection related operations and control are taken care of by transport layer. Flow and error control is also the responsibility of the transport layer.

57. **What are the major responsibilities of presentation layer? (L1)**

1. Encryption 2. Data conversion (Example: ASCII to EBCDIC) etc. are the responsibilities of presentation layer.

58. **FTP is used for file transfer and which layer owns this responsibility? (L2)**

Application layer has the ownership for file transfer. Other protocols like DHCP, SFTP, and SNMP, etc., are all application layer protocols facilitating the smooth functioning of other layers.

59. **Define the term flow control. (L2)**

It is always preferred to have acknowledgement received from the receiver for the data being sent by the sender. Flow control will confine the volume of data that the sender can propel to the receiver before waiting for acknowledgment from the receiver.

60. **Though used mostly in microprocessors, sometimes pipelining is also referred in networking. What is it? (L2)**

Before completing the processing of the current data, keeping the next set of data ready is pipelining. In networking, while sending the current set of packets, keeping the next set of packets ready to be sent is referred as pipelining.

61. **There is an equipment called brouter being used extensively in the market. What is that? (L2)**

Brouter is a device with combined features of bridge and router. It can act as a bridge and as a router whenever needed.

62. **Define the term "attenuation". (L2)**

English dictionary says it a weakening. Same will hold good for computer networks. When a signal is sent over the network, it will get weakened due to various reasons and is referred as attenuation.

**63. What are the available standards for fast Ethernet and what are the corresponding speeds?**

| Ethernet 802.3 | Speed supported | Segment (in metres) | Cable used |
|---|---|---|---|
| 10BASE5 | 10 Mbps | 500 | Co-axial |
| 10BASE2 | 10 Mbps | 185 | Co-axial |
| 10BASET | 10 Mbps | 90 | Twisted pair |

**Fast Ethernet standards and speeds.**

**64. What is the purpose of FCS in the Ethernet frame? (L2)**

A 4-byte field where a 32-bit cyclic redundancy check (CRC) value is calculated before transmission by the sender; and will be rechecked by the receiver to find out the damaged frames. If the CRC value doesn't match with the sender's FCS, the receiver will discard the transmission.

**65. What is RS in RS232? (L1)**

Recommended standard.

**66. What is the drawback with RS-232? (L2)**

The only drawback with RS-232 is the shorter distance it supports. The supported distance is only 15 metres.

**67. What is VLAN? How is it useful? (L2)**

VLAN is virtual local area network and it works by it allows dividing a LAN segment into different broadcast domains. In a normal LAN connection, computers can usually communicate using switches and routers. But, in a VLAN, logical subsets of ports are assigned on a single or multiple switches; which can control the traffic without interfering in other network channels. This improves quick rendering of data and VLAN can be used widely allowing diverse systems to communicate.

**68. Expand STP and enumerate it. (L2)**

STP is acronymic for spanning tree protocol. This is a protocol that prevents broadcast storms and formation of loops on a network. This layer 2 protocol runs on bridges and switches. Working is based on the spanning tree algorithm, shutting off all unwanted links and ports which may bring redundant loops in the network. This is the earliest of the protocols in STP series. This protocol is followed by RSTP and then MSTP.

**69. How is RSTP different from STP? (L3)**

| RSTP | STP |
|---|---|
| Its reconvergence time is lesser when compared to STP. RSTP can get regathered in less than 10 seconds. | STP needs about 30 to 50 seconds to regather. |
| RSTP has three states, namely, discarding, learning and forwarding. | STP has five states: listening, blocking, learning, forwarding and disabled. RSTP has combined three of these states—listening, blocking, and disabled—into a single port state: discarding. |
| RSTP enabled bridges can send out BPDUs with every hello time with current information. | In STP the bridges can send out a BPDU only when one is received on their ROOT port. |
| Alternate and backup port roles are introduced as two additional roles when compared to STP. | |

**70. What is a BPDU? Expand it. (L2)**

BPDU is abbreviation for bridge protocol data unit. Working with an STP algorithm, two switches can communicate data messages to each other. BPDUs are simply the self-advertisers for the switches. A switch when sending a BPDU out from it, it sends out the configuration information to the other switches in the network. BPDUs are source, through which the topology changes, link failures, loops formation are all advertised to other nodes.

**71. What is Bluetooth and what are the threats associated with it? (L2)**

Bluetooth is a short range radio frequency wireless communication standard for connecting devices in a definite perimeter which is to an extent secured by pairing. It is mostly used in portable devices as mobile phones, PDAs, tablets, laptops, headsets, watches.

There are few threats associated with Bluetooth usage and they are:

- Bluejacking: It is a message sent anonymously from one Bluetooth device to another one in the range wherein the sender doesn't know the receiver.
- Bluebugging: It will allow the hacker to take full control of the victim's equipment and without his knowledge, his/her equipment can be accessed. Even, calls or messages can be sent from the victim's equipment without his knowledge.

**72. What are the primary differences between IP Version 4 and Version 6? Which one is more beneficial? (L2)**

IPv6 is more beneficial since it can provide IPs of vast range. We are running out of IPs with IPV4 and hence is the necessity raised for IPV6.

**73. TTL is an important field in IP protocol. What is it meant for? (L2)**

TTL stands for time to live. It is an 8-bit field in an IP packet indicating the time that IP datagram should survive. It indicates the router (or whichever device, for that matter) whether a packet has expired while in the network. If it stays there for too long, the packet will be abandoned. There will be an initial time set and that will be decremented by 1 and router will be having an eye on it. When it reaches zero the datagram can be discarded.

**74. How can someone test the connectivity to some other machine in the network? (L2)**

This can be done using keyword 'ping'. One can use command prompt to use the command ping in Windows. In Linux, it can be done in terminal. Ping < IP address> is the format to be used.

**75. Where is IP address 127.0.0.1 used? How is it helpful? (L2)**

IP address 127.0.0.1 is used as loop back address. This means that it is used by the host computer to send a message back to itself. It is commonly used in troubleshooting and checking if network is functioning fine.

**76. How is ARP different from RARP? (L2)**

RARP (reverse address resolution protocol) rather an obsolete protocol, is the reverse for ARP. In ARP, MAC address was found by using IP address. In RARP, IP address can be determined by a client with the available MAC address of the device. RARP is used for diskless computers to determine their IP address.

**77. DHCP and BOOTP are the same, TRUE or FALSE. Justify. (L2)**

FALSE. DHCP is an updated version of BOOTP. BOOTP has maximum lease time of 30 days over IP addresses, whereas in the case of DHCP it is 8 days. DHCP has automatic rebinding facility, BOOTP does not have that feature, and it requires a restart. BOOTP will provide a restricted number of client IPs and it is vendor specific, while DHCP has optional parameters which support larger array of clients.

**78. What is the purpose of OSPF? (L2)**

Open shortest path first (OSPF) is the most popular layer 3 routing protocol. As the name suggests, this protocol paves the shortest way for the packets to flow when the system is too large and comprises many devices. This protocol is categorized as an interior gateway protocol (IGP). IGP protocols run within the same autonomous system. Autonomous system (AS) is a network under the same administration domain. OSPF protocol will detect the topology changes (link failures) quickly and re-converge on a new loop-free routing structures within seconds. Shortest path is computed using a method based on Dijkstra's algorithm, the shortest path first algorithm.

**79. Where can OSPF be deployed? (L2)**

OSPF can be employed on the following router types:

- Internal router
- Area border router
- Backbone router
- AS boundary routers

**80. How is BGP useful in layer 3? (L2)**

Also a routing protocol, operating on the whole WWW, BGP performs inter-domain routing in IP networks. BGP is considered to be one of the most active exterior gateway protocols (EGP), which means that it performs routing between multiple autonomous systems and exchanges reachability and routing information with other BGP systems, IPV6 and VPNs.

**81. What are the Internet control protocols? How are they useful? (L3)**

When the data is sent from source to destination (remote), IP datagram has to travel through many routers. So, there is always a chance for errors in this data transfer. Internet Control Protocol (ICMP) is the one, which will support to detect these errors. It is used in troubleshooting and diagnosis and also sending error messages indicating that a service is broken or device could not be contacted.

**82. What are the parameters that can hit quality of service? (L2)**

The following are the parameters that hit quality of service:

- Over-provisioning
- Buffering
- Congestion

**83. What will cause congestion and what are the ways to avoid it? (L2)**

Network congestion is said to have occurred when a link is carrying so much traffic in such a way that its quality of service is getting hit badly. One of the very important reasons for congestion is that traffic is often burst. If the hosts can control the rate of transmission and if it could be made more uniform, then congestion can be reduced to a great extent. This process is given the name 'traffic shaping'. Many algorithms such as leaky bucket algorithm, token bucket algorithm etc.., are used to avoid congestion.

**84. How is leaky bucket algorithm different from token bucket algorithm? (L2)**

Leaky bucket algorithm does not deal with any tokens whereas token bucket algorithm is based on tokens. Tokens are the basic parameters in token bucket algorithm whereas they do not have any impact with leaky bucket algorithm.

**85. When will be a choke packet sent? What will be the result when a choke packet is received by a sender from a router? (L3)**

Routers do a frequent check over the network by analyzing few factors like the line utilization, congestion, etc., In the event of finding a glitch, router sends a "choke packet" to the source. The source node if co-operative (most of the cases it will be so) will listen to the packet and it will cut the sending rate by 50% and will further ignore the choke packets from the router for a fixed period of time. After the cut-off time is over it will now start looking for the choke packets. If a choke packet again arrives, the rate will be cut by half again. So obviously the congestion will be seen reduced. And in the event of no choke packet arriving, the sender host may increase the sending rate.

**86. What is load shedding? When is it preferred? (L3)**

The process by which the router discards the packets that it cannot handle is called load shedding. The figure below explains the process of load shedding. Load shedding is preferred when all other mechanisms of congestion control fail.

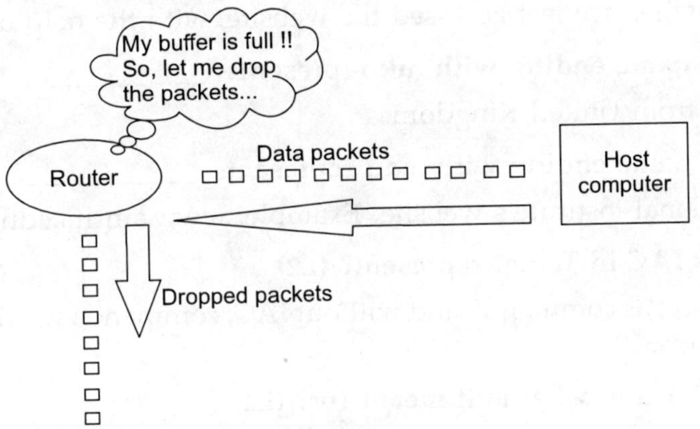


**Load Shedding.**

**87. What is jitter and how can it be reduced? (L3)**

Delay in the packets reaching the destination is termed jitter. Jitter is reduced by the routers which allocates suitable priorities to the packets, i.e., if the packet is behind schedule, the router increases the priority and sends it faster and if the packet is ahead of schedule, the router decreases the priority and schedules it for sending later.

**88. Which architecture is followed in FTP? (L2)**

File transfer protocol follows Client – Server architecture.

**89. What are the two modes of operation permitted in FTP? (L2)**

Active and passive modes.

**90. What is MIME used for and expand MIME. (L2)**

MIME is multipurpose internet mail extensions. It provides a means to interchange text in languages with different character sets.

**91. Where is POP deployed in the Email chain? What is POP? (L2)**

POP is employed at the client end to retrieve emails from remote server over TCP/IP. POP is post office protocol.

**92. How is SNMP useful in the network? (L2)**

It is used to monitor the conditions of the attached devices in the network.

**93. How are cookies useful? (L2)**

When a request is being sent to a server, the browser will search into the cookies directory to check if there is a cookie already present which is sent by that server. If present then it will also be included in the request which is about to be sent. So there will not be any need for the server to create new cookie file for the client. For instance, when we log into the bank websites, it displays a welcome message by saying when we last accessed the website, with the help of cookies.

**94. A domain name ending with .uk represents __________.**

A website from United Kingdom.

**95. A domain name ending with .edu represents __________.**

An educational institute's website. Example: www.amrita.edu

**96. What does IAC in Telnet represent? (L2)**

It is "interpret as command" and without IAC, commands would not be recognized as commands.

**97. Expand NVT and what is it meant for? (L2)**

NVT stands for network virtual terminal. Today's network is composed of heterogeneous systems and the configurations can be different from machine to machine (keyboards, character sets, resolutions, length of lines displayed) Establishing communication between these systems with TELNET is always daunting and there needs a solution. Network virtual terminal was found as a solution and it provides a standard interface.

**98. Explain the need to employ security mechanisms. (L2)**

With the growing usage of wireless technology, the threats it faces are also on the rise. Compared with wired networks, which are less prone to attacks, wireless networks are more prone to many kinds of attacks, since the medium used for communication is air! This culprit (the air) allows unauthorized people to gain access to the authorized people' conversations thereby authorized parties losing the privacy. Hence, confidentiality, integrity, privacy are the major aspects which demand the need of suitable security mechanisms.

**99. Define authorization, authentication. Differentiate between them. (L2)**

Authentication is meant to validate the identity of entities before they are allowed to participate in the network activities. Authorization must be performed sequentially to authentication mechanisms. Once an entity is authenticated, that entity has to be authorized by a third party organization to perform specific activities. In short, authentication verifies 'who you are'. Authorization checks 'what activities you are allowed to perform'.

**100. Why are ethical hackers needed? (L2)**

Ethical hackers are primarily needed to find out the loopholes in highly confidential applications where all possible security issues have to be identified and fixed.

**101. Name the command used in Linux to display the configuration details of eth0 alone? (L1)**

It can be done with the command ifconfig easily. ifconfig eth0 is the command to be used for the purpose. The following snapshot reveals the results.

```
shriramkv@shriramkv:~$ ifconfig eth0
eth0      Link encap:EthernetHWaddr f0:bf:97:18:7d:8a
inet addr:172.17.113.225 Bcast:172.17.113.255 Mask:255.255.255.0
inet6 addr: fe80::f2bf:97ff:fe18:7d8a/64 Scope:Link
          UP BROADCAST RUNNING MULTICAST MTU:1500 Metric:1
          RX packets:15377 errors:0 dropped:0 overruns:0 frame:0
          TX packets:10562 errors:0 dropped:0 overruns:0 carrier:0
collisions:0 txqueuelen:1000
          RX bytes:14616658 (14.6 MB) TX bytes:1571638 (1.5 MB)
          Interrupt:18
```

**102. How can someone enable or disable a particular interface in Linux/Unix? (L1)**

ifup or ifdown followed by the interface to be disabled or enabled is to be issued to get an interface like eth0 up or down.

**103. What is MTU? How can someone find the default MTU value in Linux? (L3)**

MTU is maximum transmission unit and it refers to the largest size of the packet which can be sent over the network. It can also refer to the frame size. The default value set for MTU is 1500 bytes and one can observe this from ifconfig eth0 command. The result will have MTU value as shown below.

```
shriramkv@shriramkv:~$ ifconfig eth0
eth0 Link encap:EthernetHWaddr f0:bf:97:18:7d:8a
inet addr:172.17.113.225 Bcast:172.17.113.255 Mask:255.255.255.0
inet6 addr: fe80::f2bf:97ff:fe18:7d8a/64 Scope:Link
          UP BROADCAST RUNNING MULTICAST MTU:1500 Metric:1
```

From the above result one can understand the MTU is set to 1500.

**104. How can someone set the MTU value other than default value in Linux/Unix machine? (L1)**

Using ifconfig command one can set the value of MTU. But, one may need administrative rights for setting this.

**105. Ping is one of the most frequent utilities in both Windows and Linux. What is abbreviated as PING? (L2)**

Packet Internet Grouper. (P IN G)

**106. Suppose one wants to ping only for 7 times, can it be done? Or it is never possible? (L2)**

Yes. It can be done with an option -c. It refers to the count and the following example reveals the same.

```
shriramkv@shriramkv:~$ ping -c 7 127.0.0.1
PING 127.0.0.1 (127.0.0.1) 56(84) bytes of data.
64 bytes from 127.0.0.1: icmp_seq=1 ttl=64 time=0.024 ms
64 bytes from 127.0.0.1: icmp_seq=2 ttl=64 time=0.036 ms
64 bytes from 127.0.0.1: icmp_seq=3 ttl=64 time=0.022 ms
64 bytes from 127.0.0.1: icmp_seq=4 ttl=64 time=0.024 ms
64 bytes from 127.0.0.1: icmp_seq=5 ttl=64 time=0.035 ms
64 bytes from 127.0.0.1: icmp_seq=6 ttl=64 time=0.035 ms
64 bytes from 127.0.0.1: icmp_seq=7 ttl=64 time=0.031 ms

— 127.0.0.1 ping statistics —
7 packets transmitted, 7 received, 0% packet loss, time 5998ms
rtt min/avg/max/mdev = 0.022/0.029/0.036/0.008 ms
```

One can see that 7 packets got transmitted and 7 is the count.

```
shriramkv@shriramkv:~$
```

**107. Traceroute command is used as "tracert" in Windows and how is it used in Linux? (L2)**

The working of the command remains the same and it does not get altered. But, the way the command is used, is to be changed. It is "traceroute". The syntax is presented below.

```
shriramkv@shriramkv:~$ traceroute www.google.co.in
```

**108. How can someone find the IP address of a website through a command in Linux? (L2)**

With the command host, one would be able to find the IP address having the website address. Following snapshot reveals the way it has been done.

```
shriramkv@shriramkv:~$ host www.amrita.edu
www.amrita.edu is an alias for amrita.edu.
```

```
amrita.edu has address 103.10.24.196
shriramkv@shriramkv:~$
```

**109. What is the command used to find out the name of the machine you are currently working on? (L2)**

hostname is the command used to achieve the task. Similarly, whoami can also be used. Following are the examples used to reveal the results.

```
shriramkv@shriramkv:~$ hostname
shriramkv
shriramkv@shriramkv:~$ whoami
shriramkv
```

**110. What does the following command do? (L2)**

tcpdump -c 7 -i eth0

It captures packets for specified interface, 7 times, i.e., 7 packets are captured.

**111. How can the command DIG used in Linux? (L2)**

Domain information grouper is the expansion of the command DIG. It is an advanced command which can render details about the hostname which is given as the part of the command. Sample results are presented below. Someone can find out the hostname with IP in hand through the same command dig -x "IP_ADDRESS".

```
shriramkv@shriramkv:~$ dig www.amrita.edu
; <<>>DiG 9.9.5-3-Ubuntu <<>> www.amrita.edu
;; global options: +cmd
;; Got answer:
;; ->>HEADER<<- opcode: QUERY, status: NOERROR, id: 15116
;; flags: qrrdra; QUERY: 1, ANSWER: 2, AUTHORITY: 0, ADDITIONAL: 1

;; OPT PSEUDOSECTION:
; EDNS: version: 0, flags:;udp: 4000
;; QUESTION SECTION:
;www.amrita.edu. IN A

;; ANSWER SECTION:
www.amrita.edu. 77218 IN CNAME amrita.edu.
amrita.edu. 77218 IN A 103.10.24.196

;; Query time: 1 msec
;; SERVER: 127.0.1.1#53(127.0.1.1)
```

;; WHEN: Thu Jun 16 11:11:59 IST 2016
;; MSG SIZE rcvd: 73

**112. How can someone get to the real time (runtime) update about all the processes running in the machine? (L2)**

Top is the command used for this purpose. It will display the details of all the processes dynamically while getting updated on the fly. But, this process itself consumes more energy than all the processes which are running in the machine. Hence, it is preferred not to use it. System admins will use jobs, bg, fg commands to know details on the processes running in the machine.

**113. If there is an unwanted process running in the machine, how can someone forcefully stop it? (L2)**

In windows, there is a well known option of CTRL + ALT + DEL which will enable user to identify the process of their choice to be killed. In Linux, one would use kill –9 <process_id> to kill a particular process. Process ID in Linux machine can be known when the process is executed with & option. It will display a number which is referred as process ID.

**114. Other than telnet how can someone login to remote machine in a secured way? (L2)**

Ssh is the abbreviation of secure shell. ssh<ip_address> would enable the user to login with username and password into a remote machine, but, in a secured way. FTP is made more secured these days through secure FTP, SFTP protocol.